Wireless Networks

Series editor
Xuemin Sherman Shen
University of Waterloo, Waterloo, ON, Canada

The purpose of Springer's Wireless Networks book series is to establish the state of the art and set the course for future research and development in wireless communication networks. The scope of this series includes not only all aspects of wireless networks (including cellular networks, WiFi, sensor networks, and vehicular networks), but related areas such as cloud computing and big data. The series serves as a central source of references for wireless networks research and development. It aims to publish thorough and cohesive overviews on specific topics in wireless networks, as well as works that are larger in scope than survey articles and that contain more detailed background information. The series also provides coverage of advanced and timely topics worthy of monographs, contributed volumes, textbooks and handbooks.

** Indexing: Wireless Networks is indexed in EBSCO databases and DPLB **

More information about this series at http://www.springer.com/series/14180

Beibei Li • Rongxing Lu • Gaoxi Xiao

Detection of False Data Injection Attacks in Smart Grid Cyber-Physical Systems

 Springer

Beibei Li 🆔
College of Cybersecurity
Sichuan University
Chengdu, Sichuan, China

Rongxing Lu 🆔
Faculty of Computer Science
University of New Brunswick
Fredericton, NB, Canada

Gaoxi Xiao 🆔
School of Electrical and Electronic
Engineering
Nanyang Technological University
Singapore, Singapore

ISSN 2366-1186   ISSN 2366-1445   (electronic)
Wireless Networks
ISBN 978-3-030-58674-4   ISBN 978-3-030-58672-0   (eBook)
https://doi.org/10.1007/978-3-030-58672-0

This Springer imprint is published by the registered company Springer Nature Switzerland AG
The registered company address is: Gewerbestrasse 11, 6330 Cham, Switzerland

This book is wholeheartedly dedicated to my respectable supervisors, as well as our group members, with whom we have worked over the years and have made it possible to reach this moment.

To my beloved families.

Preface

Building an automated, green, and efficient smart grid cyber-physical system (CPS) while ensuring high reliability and security is an extraordinarily challenging task, particularly in the ever-evolving cyber threat landscape. This challenge is also compounded by the increasing pervasiveness of information and communications technologies across the power infrastructure, as well as the growing availability of advanced hacking tools in the hacker community. One of the most critical security threats in smart grid CPSs lies in the high-profile false data injection (FDI) attacks, where attackers attempt to inject either fabricated measurement data to mislead power grid state estimation and bad data detection or tampered command data to misguide power management and control. Accordingly, FDI attacks can be subdivided into false measurement data injection (FmDI) attacks and false command data injection (FcDI) attacks.

Detection techniques for FDI attacks have been a significant research focus for smart grid CPSs to withstand these security threats and further protect the power infrastructure. However, conventional state estimation based bad data detection approaches have been proved vulnerable to the evolving FDI attacks. To meet this gap, this monograph introduces four creative research works to analyze and detect FDI attacks in smart grid CPSs.

First, a stochastic Petri net based analytical model is developed to evaluate and analyze the system reliability of smart grid CPSs, particularly against topology attacks under system countermeasures (such as intrusion detection systems and malfunction recovery techniques) in place. Evolved from FmDI attacks, topology attacks inject false data by tempering with both measurement data and grid topology information. This analytical model is featured by bolstering both transient and steady-state analysis of system reliability.

Second, a distributed host-based collaborative detection scheme is proposed to detect FmDI attacks in smart grid CPSs. It is considered in this chapter that phasor measurement units (PMUs), deployed to measure the operating status of power grids, can be compromised by FmDI attackers. Trusted host monitors (HMs) are assigned to each PMU to monitor and assess PMUs' behaviors. Neighboring HMs make use of the majority voting algorithm based on a set of predefined normal behavior rules to identify the existence of abnormal measurement data collected by PMUs. In addition, an innovative reputation system with an adaptive reputation updating algorithm is designed to evaluate the overall operating status of PMUs, by which FmDI attacks, as well as the attackers, can be distinctly observed.

Third, a Dirichlet-based detection scheme for FcDI attacks in hierarchical smart grid CPSs is proposed. In the future hierarchical paradigm of a smart grid CPS, it is considered that the decentralized LAs responsible for local management and control can be compromised by FcDI attackers. By issuing fake or biased commands, the attackers anticipate manipulating the regional electricity prices with the purpose of illicit financial gains. The proposed scheme builds a Dirichlet-based probabilistic model to assess the reputation levels of LAs. This probabilistic model, used in conjunction with a designed adaptive reputation incentive mechanism, enables quick and efficient detection of FcDI attacks as well as the attackers.

Lastly, we systematically explore the feasibility and limitations of detecting FmDI attacks in smart grid CPSs using distributed flexible AC transmission system (D-FACTS) devices. Recent studies have investigated the possibilities of proactively detecting FmDI attacks on smart grid CPSs by using D-FACTS devices. We term this approach as proactive false data detection (PFDD). In this chapter, the feasibility of employing PFDD approach to detect FmDI attacks is investigated under single-bus, uncoordinated multiple-bus, and coordinated multiple-bus FmDI attacks. It is proved that the PFDD approach is capable of detecting all these three types of FmDI attacks targeted on buses or super-buses with degrees larger than 1, as long as the deployment of D-FACTS devices covers branches forming at least a spanning tree of the power grid graph. Then, the minimum efforts demanded for activating D-FACTS devices to detect the considered three types of FmDI attacks are evaluated. In addition, the limitations of PFDD are also discussed, and it is strictly proven that this approach is not able to identify FmDI attacks on smart grid CPSs that are targeted on buses or super-buses with degrees equaling 1.

Chengdu, Sichuan, China   Beibei Li
Fredericton, NB, Canada   Rongxing Lu
Singapore, Singapore   Gaoxi Xiao
June 2020

Acknowledgements

There have been many people who have walked alongside me during my Ph.D. journey. They have guided, supported, and accompanied me. I would like to, hereby, thank each and every one of them sincerely.

First and foremost, I would like to express my deepest gratitude to my respectable supervisors—Dr. Xiao Gaoxi at Nanyang Technological University (NTU), Singapore, and Dr. Lu Rongxing at the University of New Brunswick (UNB), Canada— for their unwavering support and constructive guidance throughout this thesis. They, upon whose shoulders I stand, explored and paved the path before me. Without them, this thesis would simply not have been possible. Such academic rigor as may be found in this thesis is largely due to Dr. Xiao's refusal to let me get away with things, while his unerring sense of when and how to intervene has taught me not only as a good researcher but also a potential good tutor. He is always willing to take time to listen and usually provide insightful questions and comments, as well as clear instructions as feedback. Dr. Lu is a renowned expert in cybersecurity domain, whose passion for doing research and teaching has set a new standard for everyone involved. His unstinting support and encouragement have driven me to strive for excellence. Having also a friend figure, Dr. Lu is really a nice guy who cares about his students not only on their research career but also on daily lives. Many thanks are also due to Dr. Wang Licheng at Beijing University of Posts and Telecommunications (BUPT), China, who started me down this road with selfless support, encouragement, and guidance.

I would especially acknowledge my Thesis Advisory Committee (TAC) members—Dr. Zhang Jie and Dr. Ma Maode at NTU. Thanks for their faith in my ability and continuous support ever since I joined NTU. I hope this research opens up opportunities for us to do research together in the future.

I would extend my heartfelt gratitude to Dr. Ali A. Ghorbani, Dr. Kim-Kwang Raymond Choo, Dr. Bao Haiyong, Dr. Deng Ruilong, Dr. Wang Wei, and Dr. Luo Sheng, who contributed to the making of this monograph. Thanks for their constructive criticism, which enabled me to improve my research and writing skills. Particular thanks must also be recorded to Dr. Xu Chang, Dr. Liu Yali, Dr. Kong Qinglei, Dr. Zhao Ming, Dr. Meng Min, Dr. Lin Changlu, Dr. Liu Ximeng, Dr. Li

Chen, Dr. Li Lichun, Dr. Hu Hao, Dr. Zhai Chao, Mr. Huang Cheng, Mr. Wang Guoming, Mr. Katuwal Rakesh, Mr. Cheng Shuo, Mr. Hao Changyu, Mr. Zhang Hehong, and Mr. Li Xiang who offered collegial guidance and support over the years.

There are many of my friends to name individually; however, special thanks are given to Dr. Yang Rong, Dr. Li Dan, Dr. Yang Ming, Dr. Bi Hui, Dr. Ma Lijia, Dr. Zhang Heng, Dr. Wang Zeng, Dr. Chen Chunyang, Ms. Sun Meng, Ms. Gao Yumeng, Ms. Gong Bo, Ms. Huang Yi, Ms. Wang Zhenzhen, Ms. Huang Rui, Ms. Xin Jian, Ms. Chen Qian, Ms. Wang Yongheng, Ms. Li Yanan, Ms. Chen Shi, Ms. Chen Qiyin, Ms. Zhang Lili, Ms. Ouyang Qingling, Ms. Zhang Ran, Mr. Cheng Yanyu, Mr. Liu Yunxiang, Mr. Pan Zihan, Mr. Wang Yang, Mr. Shao Hongxin, Mr. Li Xiaochen, Mr. Zhang Yunlong, Mr. Guo Zhihong, Mr. Zhang Songze, and Mr. Gao Li at NTU and Dr. Yang Haomiao, Dr. Huang Junlin, Ms. Yang Xue, Ms. Zheng Yandong, Ms. Li Huixia, Ms. Deepigha S. V. Babu, Mr. Guo Wei, Mr. Xu Chenghao, Mr. Hassan Mahdikhani, Mr. Saeed Shafeiee Hasanabadi, Mr. Tao Xi, Mr. Zhang Xichen, Mr. Xiao Hongtao, and Mr. Zhang Yongcan at UNB. Thanks for giving me a wonderful and memorable life at both NTU and UNB. In addition, my wholehearted thanks are given to Mrs. Han Yuhong, Ms. Xie Wanlun, Ms. Yu Pan, Mr. Du Wuhang, and Mr. Chen Dawen at BUPT for their generous support, understanding, and encouragement given in many moments of crisis over the years. I cannot list all the names here, but you guys hold a special place in my heart.

Finally, and most importantly, my most heartfelt and forever gratitude goes to my family and girlfriend, who have always been a constant source of support and encouragement. Thanks to my parents and elder sister for putting me through the best education possible and giving me the strength to reach for the stars and chase my dreams. I appreciate their sacrifices and unending support, and I would not have been able to get to this stage without them.

Contents

Acronyms

AC	Alternating current
CC	Control center
CPS	Cyber-physical system
DC	Direct current
DDoS	Distributed denial-of-service
DoS	Denial-of-service
D-FACTS	Distributed flexible AC transmission system
FcDI	False command data injection
FDD	False data detection
FDI	False data injection
FmDI	False measurement data injection
GPS	Global positioning system
HM	Host monitor
HMI	Human–machine interface
ICS	Industrial control system
ICT	Information and communications technology
IDS	Intrusion detection system
IED	Intelligent electronic device
IoT	Internet-of-things
IPS	Intrusion prevention system
LA	Local agent
LAN	Local area network
LMP	Locational marginal price
MLE	Maximum likelihood estimation
MTTD	Mean time to disturbance
MTTF	Mean time to failure
MTTM	Mean time to malfunction
MxST	Maximum spanning tree
PDC	Phasor data concentrator
PLC	Programmable logic controller
PMU	Phasor measurement unit

RF	Radio frequency
RTU	Remote terminal unit
SCADA	Supervisory control and data acquisition
SDH	Synchronous digital hierarchy
SONET	Synchronous optical networking
SPN	Stochastic Petri net
WAMS	Wide area measurement system
WAN	Wide area network

Chapter 1
Introduction

1.1 Background

With energy being a premium resource for economy, society, and national security, ensuring an accurate, reliable, and efficient power generation, transmission,
distribution, and consumption is of prime concern in the twenty-first century
[1]. Unfortunately, the massive blackout in northeastern North America in 2003
uncovered the ease with which the electric grids could be taken down. It is, indeed,
not the end of the story for the security and reliability breaches of existing electric
grids. To build a fully automated, resilient, and self-healing smart grid, a series
of advanced technologies including information and communications technologies
(ICTs), automation, distributed control, wide area monitoring and control, smart
metering, to name a few, are rapidly incorporated into the existing electric grid over
the recent years [2, 3]. Due to lack of strong and diligent security measures in place,
however, these newly introduced technologies—exposing a great number of access
points to the public—have been opening up possibilities for malignant penetrations
[4].

1.1.1 Cybersecurity Events Relating to Electric Grids

Cyberattacks on electric grids are no longer a theoretical concern. The summer of
2010 stroke the world in an unprecedented way by discovering the world's first
digital weapon—Stuxnet [5]. Unlike any other computer virus or worm that came
before, Stuxnet escaped the digital realm to wreak physical destructions on the
equipment that computers controlled. By infiltrating the Windows computers at the
Natanz nuclear plant in Iran, Stuxnet destroyed an estimate number of 984 uranium
enriching centrifuges in total [6]. The impacts of this event go beyond the immense

© Springer Nature Switzerland AG 2020
B. Li et al., *Detection of False Data Injection Attacks in Smart Grid
Cyber-Physical Systems*, Wireless Networks,
https://doi.org/10.1007/978-3-030-58672-0_1

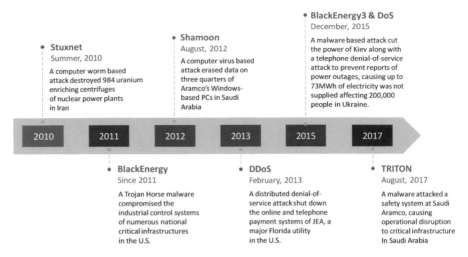

Fig. 1.1 A timeline of recently reported significant cybersecurity events on electric grids

damages caused to Iran. A great deal of ideas of copying and re-purposing Stuxnet from the hacking community as well as, correspondingly, research studies focusing on detection and mitigation of such cyberattacks from the academia and industry have quickly emerged thereafter.

Started by Stuxnet, a huge wave of cybersecurity events on electric grids have been observed since then (see Fig. 1.1 a timeline of these events).

Stuxnet	"Stuxnet" is a virus sweeping the world of industry. The worm was specifically designed to destroy SCADA equipment used by Iran in its nuclear fuel enrichment process. This attack successfully damaged SCADA equipment in many places in Iran. Although nations will resort to mutual cyberattacks before 2010, "Stuxnet" was the first cybersecurity event that shocked the world, from the theft of single information data to the destruction of actual physical facilities, which marks the beginning of a new phase in cyber warfare.
BlackEnergy	It is reported in 2011 that a cyber campaign involving a Trojan Horse based malware, also notorious as BlackEnergy, compromised the industrial control systems (ICSs) of numerous national critical infrastructures in the USA [7]. The BlackEnergy campaign used previously unknown software vulnerabilities in multiple common Human Machine Interface (HMI) software products to gain direct access to control system operating screens. The malicious code could potentially be used to manipulate control processes and cause physical damage. No

Shamoon

DDoS

BlackEnergy3 & DoS

TRITON

interaction with the target was required as BlackEnergy targeted systems connected directly to the Internet.

Shamoon is a new type of virus similar to the advanced persistent threat (APT) attack viruses like Flame, which target energy companies or energy departments and permanently erase data from infected Windows machines. In August 2012, a self-replicating computer virus named Shamoon infected three-quarters of Windows-based corporate PCs at Saudi Aramco, one of the world's largest oil companies [8].

Distributed denial-of-service (DDoS) attack refers to multiple attackers in different positions attacking one or more targets simultaneously, or an attacker could take control of multiple machines in different locations and use them to attack the victim at the same time. Since the point of origin of an attack is located in different places, this type of attack is called a DDoS attack, in which the attacker can have more than one. In February 2013, JEA, the seventh-largest community-owned electric utility in the USA was hit by a DDoS attack, which led to a crash of online and telephone payment systems for a few days [9].

The year 2015 has witnessed the world's most sophisticated and most successful cybersecurity event in electric grids to date [10]. It was a Saturday night just two days before Christmas in 2015, an orchestrated cyberattack simultaneously hijacked several power distribution centers at the Ivano-Frankivsk region of Western Ukraine. Approximately 30 substations were eventually taken offline in this assault, leaving more than 230,000 Ukrainians in the dark for a period of one to six hours.

This assault was launched in a well-choreographed dance, where well-trained hackers synchronously switched off a number of substations, disabled IT infrastructures, destroyed files stored on the servers, as well as initiated a telephone denial-of-service (DoS) to deny customers' reports of power blackouts. It was exactly the "brilliant" plan for launching such a real cyberattack that shaped the research landscape of both the power and security community.

An analogous attack on Saudi Aramco was initiated in August 2017, where a malware called TRITON created operational disruptions towards critical infrastructures in Saudi Arabia [11]. TRITON is malware designed to target Safety instrumented system (SIS), whose attack on the SIS controller is extremely dangerous. Once the controller

Nation	Actions and Investments
The U.S.	Since 2010, the U.S. has invested more than $210 million in cybersecurity research, including advancing the resilience of the Nation's energy delivery systems; The American Recovery and Reinvestment Act prompted more than $4.5 billion investments in grid modernization involving improving the grid reliability and resilience; The Joint United States-Canada Electric Grid Security and Resilience Strategy was issued in 2016 to promote the development of a security and resilience strengthened North American electricity grid.
Canada	An investment of estimated CAD $25-40 billion by 2020 was announced to refurbish, rebuild, replace the grid infrastructure to ensure a reliable power transmission system; In January, Canada's Minister of Natural Resources announced a $100-million call for proposals to fund more smart grid systems to fight climate change, create clean jobs and ensure safer power delivery for Canadians; The Joint United States-Canada Electric Grid Security and Resilience Strategy was issued in 2016 to promote the development of a security and resilience strengthened North American electricity grid.
China	A total amount of RMB 286.11 billion was invested in upgrading smart substations and smart meters during the 12-th Five-Year Plan (2011-2015); At least RMB 2 trillion will be spent to improve the reliability of power transmission over the 13-th Five-Year Plan (2016-2020).
The U.K.	Up to £6.1 billion of electricity network investment in Scotland and around £11.6 billion in England and Wales were approved in 2013 to build reliable power delivery networks.

Fig. 1.2 A summary of recent significant actions and investments in security and reliability of electric grids

is breached, the hacker can reprogram the device to trigger the security state and have a significant impact on the operation of the target environment.

1.1.2   Recent Actions and Investments in Security and Reliability of Electric Grids

The frequent and horrible cyberattacks in recent years have emerged as a driving factor to promote the advancements of existing electric grids. It becomes a common sense for nation governors that a secure and reliable power delivery network is of utmost importance to support a functioning society. Mindful of this, they have been making action plans and directing investments to reinforce the security and reliability of their electric grids.

Figure 1.2 summarizes several recent significant actions and investments in security and reliability of electric grids in the USA, Canada, China, and the UK [12–16]. As we see, each nation has announced enormous investments as well as necessary joint strategies to improve the electric grids' security and reliability.

The USA   Since 2010, the USA has invested more than $210 million in cybersecurity research including advancing the resilience of the Nation's energy delivery systems;

The American Recovery and Reinvestment Act prompted more than $4.5 billion investments in grid modernization involving improving the grid reliability and resilience;

	The Joint United States-Canada Electric Grid Security and Resilience Strategy was issued in 2016 to promote the development of a security and resilience strengthened North American electricity grid.
Canada	An investment of estimated CAD \$25–40 billion by 2020 was announced to refurbish, rebuild, replace the grid infrastructure to ensure a reliable power transmission system; In January, Canada's Minister of Natural resources announced a \$100-million call for proposals to fund more smart grid systems to fight climate change, create clean jobs, and ensure safer power delivery for Canadians; The Joint United States-Canada Electric Grid Security and Resilience Strategy was issued in 2016 to promote the development of a security and resilience strengthened North American electricity grid.
China	A total amount of RMB 286.11 billion was invested in upgrading smart substations and SMs during the 12-th Five-Year Plan (2011–2015); At least RMB 2 trillion will be spent to improve the reliability of power transmission over the 13-th Five-Year Plan (2016–2020).
The UK	Up to £6.1 billion of electricity network investment in Scotland and around £11.6 billion in England and Wales were approved in 2013 to build reliable power delivery networks.

1.2 Brief Introduction of a Smart Grid CPS

With such a research background in mind, in this section, we brief the concepts and architectures of cyber-physical systems (CPSs), the smart grid CPS, the supervisory control and data acquisition (SCADA) system, as well as the wider area measurement system (WAMS), respectively.

1.2.1 Cyber-Physical System

A CPS is an integrated, hybrid network of cyber and engineered physical elements. It is co-designed and co-engineered by experts from various domains, including control & automation, computer science, communications, mechanics, etc., to create an adaptive, flexible, situation aware, and predictive hybrid system. Through the organic integration and deep collaboration of $3C$ (Computer, Communication, and Control) technologies, real-time perception, dynamic control, and information service of large-scale engineering system can be realized. CPS can realize the integrated design of computing, communication, and physical system, which can make the system more reliable, efficient, and real-time cooperation, and has an important and extensive application prospect. Information physical system includes the ubiquitous environment perception, embedded computing, network communi-

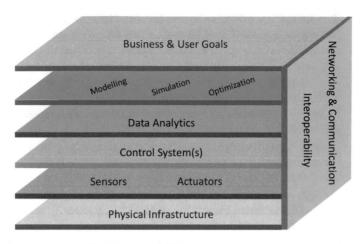

Fig. 1.3 A common layered architecture of CPSs

cation, and network control system engineering in the future. It pays attention to the close combination and coordination of computing resources and physical resources and is mainly used in some intelligent systems, such as device interconnection, IoT sensing, smart home, robot and intelligent navigation.

A common layered architecture of CPSs are presented in Fig. 1.3. In this reference architecture, the bottom layer is the large-scale physical infrastructure. Widespread sensors and actuators are deployed over the physical infrastructure to measure its operating status and execute given control commands towards it. Data collected by sensors will be reported to the control systems, who also issue the control commands to the actuators. On top of the control systems, data analytic techniques will be employed to analyze these reported data to further support various applications such as system modeling, simulation, and optimization. Owners of the CPSs will make use of these invaluable data as well as corresponding applications for business purposes and user goals. Note that the networking & communication technologies running throughout all the layers making its interoperability possible.

1.2.2   Smart Grid CPS

A smart grid CPS, grounded on the physical electric grid, incorporates advanced digital technologies, automation, computer and control to perform a duplex two-way communications between the customers and utilities. Also, a smart grid CPS can be regarded as an Internet-of-Things—power generators, distributors, meters, utilities, and customers. It is expected that by employing two-way communications, a smart grid CPS cannot only enable monitoring and controlling of power delivery in a (near) real-time mode, but also allow customer interactions of electricity usage.

The promising benefits of the smart grid CPS include [17]

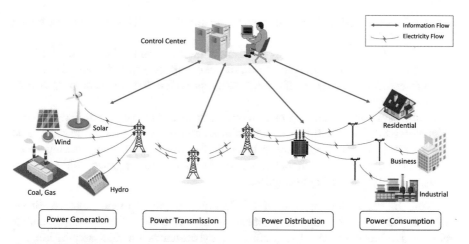

Fig. 1.4 The architecture of a smart grid CPS

- improved power reliability and quality
- increased resilience against system faults or natural disasters
- auto-scheduling of power delivery
- predictive maintenance, self-healing, and fast remote repair
- increased capacity and energy efficiency, and reduced carbon emissions
- expanded integration of renewable energy sources, e.g., wind, solar, hydro
- load shedding and lowered electricity tariff
- enhanced customer knowledge of energy usage
- real-time pricing
- remote billing and reduced of manpower costs
- support for smart cities and intelligent transportation systems

Figure 1.4 shows the architecture of a smart grid CPS, with which the above-mentioned promising benefits can be achieved. A smart grid CPS mainly consists of power generation, power transmission, power distribution, power consumption, and control systems.

- *Power generation*: Energy resources can be classified into renewable (e.g., wind, solar, hydro, biomass, geothermal) and non-renewable resources (e.g., coal, gas, nuclear). Power generation can be classified into centralized generation and distributed generation.
- *Power transmission*: Electricity is transmitted from generating points to substations through power transmission networks, usually at high voltages—115kV and above—to reduce power loss over long-distance transmission.
- *Power distribution*: Electricity is delivered from the substations to the customer premises through power distribution networks. Transformers are employed to lower the high voltage from transmission networks to medium voltage ranging

from 2kV to 35kV, and lower it again to the utilization voltage when approaching
to the customer premises.

- *Power consumption*: Electricity consumers include residential houses, business
  bodies (e.g., schools, hospitals, commercial buildings), and industrial plants.
  Differential electricity tariffs are usually provided for different purposes of
  energy usages.
- *Control systems*: In existing electric grids, there is usually a centralized control
  center, while there will be more distributed control centers in future smart grid
  CPSs. The control centers are responsible for power monitoring, management,
  and control to ensure a reliable, secure, and efficient power generation, transmission, distribution, and consumption.

With the rapid development of the Internet, network threats against circuit
transmission networks, especially attacks against SCADA systems, have become
more and more common. The number of cyberattacks on data acquisition and
Surveillance control (SCADA) systems nearly doubled in 2015, and the number
of cyberattacks on SCADA systems increased nearly 600% from 2012, according
to a report from Dell's security division. The researchers speculate that such attacks
will get worse over the next few years.

1.2.3 The Architecture of SCADA

SCADA is an industrial computer-based control system employed to gather and
analyze the operating status data of the industrial equipment, and to further manage
and control the industrial process. The architecture of SCADA in electric grids is
presented in Fig. 1.5.

It consists of a SCADA server, a HMI, multiple numbers of remote terminal units
(RTUs), programmable logic controllers (PLCs), and communication interfaces.
Specifically,

- *SCADA server*: Serving as a centralized master, the SCADA server monitors and
  controls the whole system by analyzing the telemetry data reported from the
  RTUs and generate corresponding feedback orders. These orders will be issued
  to either RTUs or PLCs.
- *HMI*: Located in the control room, the HMI presents visualized operating
  status, scheduled maintenance procedures, logistic information, and diagnostic
  information of the whole system. In addition, it also provides the interactive
  interface for a system operator to enable human control and management.
- *RTUs*: Also termed as remote telemetry units, RTUs are located at the remote
  substation employed to gather telemetry data from field devices. RTUs also
  process simple orders from the SCADA server, e.g., orders for controlling the
  connected physical objects.
- *PLCs*: As the last-mile controllers, PLCs process the orders from the SCADA
  server to trigger a set of actions such as turning on/off a line breaker, increasing

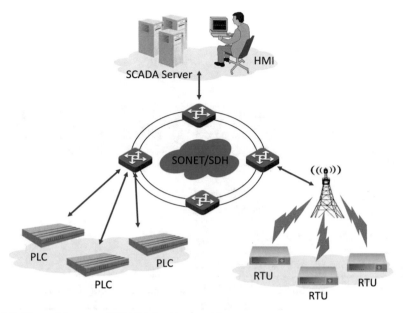

Fig. 1.5 The architecture of SCADA in electric grids

power generation. PLCs are microcomputer based devices that have advanced data handling, storage, and communication capabilities.

- *Communication interfaces*: A number of communication interfaces constitute the communication infrastructure across the large-scale SCADA system to deliver message among all entities in a SCADA system. Radio frequency (RF) and directed wired connections are usually used for local communications in a SCADA system, while synchronous optical networking (SONET) and synchronous digital hierarchy (SDH) are frequently used for backbone communications.

1.2.4 The Architecture of WAMS

WAMS, also called WAMCS (wide area measurement and control system) by some researchers, now offers a supplement to existing SCADA system. By incorporating new ICTs, WAMS is able to provide a highly accurate, dynamic, and real-timely view of electric grids. WAMS is also featured by phasor synchronization and time stamping of the system operating status data. The architecture of a WAMS is provided in Fig. 1.6. As is shown, a WAMS is comprised of a control center (CC), a set of phasor measurement units (PMUs), a set of phasor data concentrators (PDCs), global positioning system (GPS), and communication infrastructure—SONET/SDH based backbone and wired local communication networks [18]. Specifically,

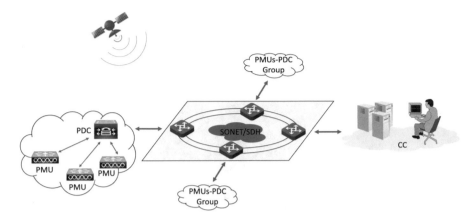

Fig. 1.6 The architecture of a WAMS

Table 1.1 The variables that PMUs can measure

State variable	Description
f	Signal frequency
Δf	Signal frequency variation rate
L_{MW}, L_{Mvar}	Load MW and load Mvar
$P/\delta_P, Q/\delta_Q$	Active and reactive power phasor
$V_A/\Theta_{V_A}, V_B/\Theta_{V_B}, V_C/\Theta_{V_C}$	Phase A, phase B, and phase C voltage phasor
$I_A/\Theta_{I_A}, I_B/\Theta_{I_B}, I_C/\Theta_{I_C}$	Phase A, phase B, and phase C current phasor
$V_1/\Theta_{V_1}, V_2/\Theta_{V_2}, V_0/\Theta_{V_0}$	Positive-, negative-, and zero-sequence voltage phasor
$I_1/\Theta_{I_1}, I_2/\Theta_{I_2}, I_0/\Theta_{I_0}$	Positive-, negative-, and zero-sequence current phasor
On/off	State of breakers

- *PMUs*: As major measurement devices, PMUs collect synchronized and time-stamped data of power system operating status such as voltage magnitude and phase, current magnitude and phase, power frequency, and change of frequency. These real-time data are collected at a usual frequency of 50/60Hz and then reported to the regional PDC via local area networks (LANs). The state variables that a PMU can measure are listed in Table 1.1.
- *PDCs*: The reported measurement data from PMUs are aggregated by regional PDCs and then sent to CC via backbone communication networks.
- *CC*: As a centralized system controller, CC is in charge of management and control of the whole electric grids by analyzing the real-time measurement data, diagnostic data, scheduled information, etc.
- *GPS*: The highly precise synchronous global positioning signals provided by GPS enable the synchronization of measurement data, which also makes it possible for CC to conduct analysis of synchronous data across the whole electric grid.

- *Communication infrastructure*: Serving as courier among all the entities in smart grid CPS, the communication infrastructure plays a significant in delivering the measurement data, commands, and any type of information across the electric grid.

1.3   Security Requirements, Challenges, and Research Motivations

In this section, we brief the security requirements that must be fulfilled to build a secure smart grid CPS, the security challenges that a smart grid CPS is facing with, and the research motivations that stimulate our research studies.

1.3.1   Security Requirements

From a cybersecurity's perspective, the security requirements can be classified into the following three categories [19]:

- **Integrity**: Protecting against the unauthorized modification or destruction of information, e.g., measurement data from meters or command data from control systems. Modified or destructed information opens the door for mishandling of information, leading to mismanagement of power or malfunction of power applications. In the field of information security, the integrity of information assets also often means accurate and correct, untampered, can only be changed by accepted methods, can only be changed by authorized personnel or process, meaningful and usable. Factors influencing information integrity are as follows: the source of information, including where, how, and through whom the information is obtained; the status of the information being protected before its arrival; the protection of information upon its arrival in the organization, etc. Integrity is a feature that network information cannot be changed without authorization. That is, the network information in the storage or transmission process is not accidentally or deliberately deleted, modified, forged, out of order, replay, insert and other damage and loss of features. Integrity is an information-oriented security that requires information to be kept as it is, that is, the information is correctly generated, stored, and transmitted. Integrity differs from confidentiality in that confidentiality requires information not to be disclosed to unauthorized persons, while integrity requires information not to be compromised for any number of reasons. The main factors affecting the integrity of network information include device fault, error code (error code generated during transmission, processing and storage, error code caused by reduced stability and accuracy of timing, error code caused by various interference sources), human attack, computer virus, and so on.

- **Confidentiality**: Protecting privacy and proprietary information, e.g., customer energy consumption data collected by smart meters (SMs), by authorizing restrictions on information access, usage, and disclosure. Confidentiality is the property of a network where information is not disclosed to or exploited by unauthorized users, entities, or processes. That is, a feature that prevents information from being leaked to unauthorized individuals or entities and that information is used only by authorized users. Confidentiality is an important means to ensure network information security on the basis of reliability and availability. The commonly used security techniques include: anti-detection and collection (to prevent useful information from being detected by the opponent), radiation prevention (to prevent useful information from radiating out in various ways), and information encryption (to encrypt information with encryption algorithm under the control of the key), physical confidentiality (using various physical methods, such as restriction, isolation, masking, control, and other measures, to protect the information from disclosure).

- **Availability**: Ensuring timely and reliable access to information and services, e.g., command data from control systems and real-time electricity price to customers. Compromised availability may cause delayed power delivery or increased electricity budgets. Availability is a feature of network information that can be accessed by authorized entities and used on demand. That is, the features that allow authorized users or entities to use network information services when needed, or that can still provide effective services to authorized users when part of the network is damaged or needs to be degraded. Availability is the user-oriented security performance of network information system. The most basic function of network information system is to provide services to users, and users' requirements are random, multi-faceted, and sometimes time requirements. Availability should also meet the following requirements: identity and confirmation, access control (control of the user's permissions, only access to the appropriate permissions of resources, to prevent or limit unauthorized access through hidden channels. Including discretionary access control and mandatory access control), the business flow control (using the method of partitioning load, to prevent excessive concentration of business flow caused by network congestion), routing control (choose subnet that is stable and reliable, trunk or link, etc.), an audit trail (of all security incidents in the network information system is stored in the security audit tracking, in order to analysis the reason, distinguish responsibility, take corresponding measures in a timely manner. The information of audit tracking mainly includes: event type, level of managed object, event time, event information, event response, and event statistics.

- **Reliability**: Reliability is the characteristic that network information system can complete the specified functions under specified conditions and within specified time. Reliability is one of the most important requirements for system security and is the construction and operation target of all network information systems. There are usually three measures for the reliability of network information system: resistance, survivability, and effectiveness. Damage resistance refers to the reliability of the system under the man-made damage. For example, if some lines or nodes fail, the system can still provide a certain level of service. Enhanced resistance to destruction can effectively avoid large areas of paralysis caused by various disasters (war, earthquake, etc.). Survivability is the reliability of a system under random failure. Survivability mainly reflects the effects of random failure and network topology on system reliability. Here, random damage refers to the natural failure of system components due to natural aging. Validity is a kind of reliability based on business performance. The effectiveness of network information system is mainly reflected in the failure of components to meet the requirements of business performance. For example, network component failures do not cause connectivity failures, but they do cause quality indicators to drop, average latency to increase, and line congestion. Reliability is mainly manifested in hardware reliability, software reliability, personnel reliability, environmental reliability, and so on. Hardware reliability is the most intuitive and common. Software reliability refers to the probability of a program running successfully within a specified period of time.

- **Authentication**: Protecting against invalid users joining proprietary computer and communication systems, e.g., control systems, to ensure the system users are authentic. Unauthenticated users may deceive or mislead control systems' decisions, or exhaust system resources.
- **Authorization**: Ensuring access to system information and services are legitimate. It is also referred as access control. Unauthenticated user may misuse or jeopardize the system resources.

1.3.2   Security Challenges

With the rapid development of smart grid technologies, its data and information security are facing growing challenges due to, for example, the increasing difficulty of information security management, the increasing complexity of security system, and the inherent vulnerabilities in existing power infrastructures, etc.

- **High complexity of smart gird**: With the increasing scale of smart grid, the power structure becomes more and more complex, and the security and stability problems of smart grid become obvious. The more complex of the smart grid is, the higher coupling degree of the smart grid, the more complex of the division of system security region, which means the security protection of the smart grid is more and more difficult.
- **Lack of smart grid security system**: In recent years, more and more attention has been paid to the security of smart grid with the development of economy. Although some developed countries have proposed a targeted strategy, the unified smart grid security system has not been completed. Because existing security systems are not yet perfect, the security of smart grid has been affected to some extent, which is also a serious problem faced by the current smart grid construction.

- **Complexity of network environment**: With the development of smart grid construction, the information integration is higher, the network environment is more and more complex, and the degree of protection against virus attack is also higher. At the same time, many information communication technologies provide great support for the management and operation of smart grid. However, some information security risk also appears in each link of smart grid operation. In order to promote the speed of data transmission, the smart grid needs the public network to transmit important data information, which brings a new threat to the security and stability of the smart grid, in serious cases, it may cause a serious safety accident. The imperfect intelligent network system makes it possible for hackers to take advantage of it, which may pose a security threat to the smart grid. The way of network attack is also gradually diversified, which will bring inestimable harm to smart grid system.
- **Difficulties in guaranteeing information security of users**: During the operation of smart grid system, the connection between the smart grid and the user becomes closer and closer, and the information and intelligence are combined and interacted. This imperceptibly increases the security risk of user information. Firstly, the information exchange between users and power enterprises needs to rely on the public network, which may expose the privacy of users' personal information data, at the same time, some household electrical equipment is exposed to the power system, which allows hackers to take advantage. Therefore, the information security of smart grid is very important.
- **Security risks in smart devices**: Smart devices can monitor the operation of the power network in real-time, locate and repair the faults accurately and effectively, which guarantees the safe operation of smart grid system. The operation of smart devices can provide some support for remote access, such as remote disconnection, software upgrade, and so on. But it would also pose a security threat to the smart grid. The hacker can invade the intelligent terminal according to the vulnerability of the software system, and control the intelligent system, so as to expose the user's record information and even control the power system. As a result, these mobile devices have the potential to be an entry point for hackers into the smart grid. With the diversification of access modes of intelligent equipment and the increasing complexity of equipment environment, more and more requirements are put forward for information security protection.

These security challenges in smart grid CPS can be classified into three categories, i.e., cyber threats, physical threats, and cyber-physical threats.

1.3.2.1 Cyber Threats

Networks in smart grid CPSs can be divided into home area network (HAN), neighborhood area network (NAN), and wide area network (WAN). Cyber threats can originate from either of these three types of networks.

Table 1.2 Some representatives of cyberattacks

Security breach type	Example attacks
Integrity	False data injections (FDIs), man-in-the-middle
Confidentiality	Eavesdropping, theft
Availability	DoS, DDoS
Authentication	Malware, Trojan
Authorization	Malware, Trojan, spoofing

- Cyber threats to HANs: Typical home domain network attacks target smart home appliances and smart electricity meters. Since these devices are easily available and inexpensive to attack, they are particularly vulnerable to malicious users. By means of reverse engineering, attackers can easily find out the communication protocol vulnerabilities and design defects in such equipment, so as to carry out large-scale attacks on the same equipment, thus affecting the security of the entire power grid.
- Cyber threats to NANs: The main targets of neighborhood network attacks are power substation and power distribution center. It is difficult for general attackers to directly enter the power substation and power distribution center to carry out attacks, but the neighborhood network and wide area network are connected through the neighborhood network gateway, and attackers can launch attacks against the communication link and communication protocol vulnerability.
- Cyber threats to WANs: The main targets of cyber threats against WANs are power stations and control equipment. These threats are the most damaging, and if successful, major security incidents may be caused, such as massive power outages.

Security threats coming from the cyber domain cause damages or compromise the operating efficiency of electric grids usually by breaching the aforementioned security requirements. Table 1.2 presents the security breach types with corresponding examples of cyberattacks.

1.3.2.2 Physical Threats

Since 2009, the Department of Homeland Security has been tracking cyberattacks on critical infrastructure, including emergency services, water treatment, and communications networks. The power sector, which includes power companies and power plants, accounted for 40% of all 887 reported attacks. All of America's 900 urban and rural electric cooperatives are non-profit, with only members and no shareholders. They serve about 12% of the nation's population but cover three-quarters of the country. Even in rural areas, they provide electricity to key service industries such as hospitals, factories, and mines. They have a smaller staff and budget to deal with cyberattacks than their counterparts in the open market, which makes them more comfortable talking about cyberattacks. In fact, they have to, because sharing information and resources is the only way for them to survive.

In the smart grid information security threats, the most common information security threat factors for smart grid equipment brought hidden dangers. Physical equipment plays an important role in the development of smart grid, which is the foundation of smart grid development. Unlike traditional power grids where technical staffs are responsible for detecting faults and dangers, smart grids employ advanced technology and equipment to automatically detect and process faults and possible dangers. Relies on the restoration of intelligence technology can only deal with simple faults encountered in the operation of the smart grid and, for the complex faults in the operation of the smart grid, although you can rely on intelligent technology to repair, but the lack of technical personnel in the process of repair, the supervision of information security problems appears easily, thus, a hidden trap network intruders can be hidden trouble to steal important information, and even lead to paralysis of the entire grid system.

The physical threats on electric grids include

- cutting fiber optic cables to shut down telecommunication lines
- destroying field equipment, e.g., transformers, cameras, sensors
- destructing documents, installations, and materials
- theft of proprietary information, or equipment
- on-site tempering of electronic devices
- field measurement and investigation

1.3.2.3 Cyber-Physical Threats

Recent years have been experiencing an increasing trend of cyber-physical threats on electric grids. Cyber-physical threats are usually orchestrated combinations of single cyber threats and physical threats, which are more complicated and threatening than either single threats. For example, cyber-physical attacks can disable cameras, turn off a building's lights, make a car veer off the road, or physically compromise and/or destroy some sensors, followed by maliciously reporting falsified sensing data on their behalf. Cyber-physical attacks are replacing physical attacks in crime, warfare, and terrorism, causing great harm to the whole world.

False data injection (FDI) attack is a typical cyber-physical attack. This attack can mislead the state estimation by altering the data accessed by remote devices, intercepting the traffic flow by malicious means, or transmitting the data through the Cooperative Operation Network, so that the SCADA system can transmit the altered data, bypassing the traditional bad data detection (BBD) mechanism without affecting the data integrity, the Control Center will issue wrong control instructions and control operations to the physical equipment, resulting in equipment failure or stop running. The results may block transmission lines, affect the functions of event analysis, optimal power flow, economic dispatch, and disrupt the effectiveness of distribution process, resulting in large amounts of energy supply losses, which increases the energy transmission cost and the number of power users, damages the operation of real-time power market, causing serious physical device damage, economic loss, and even casualties. Section 1.1.1 mentioned an attack on Ukraine's State Grid that resulted in widespread power outages and severe economic losses. This event is known as the milestone event of the smart grid cyber-physical attack, which shows that FDI has moved from a theoretical model to a real-world one. Depending on the target, we can classify it into four categories: device attacks, data attacks, privacy attacks, and network accessibility attacks.

1. **Device attacks:** Aimed to compromise network terminal devices for further attacks, such as data attacks and network accessibility attacks. These devices cover the entire smart grid, including on-site devices that collect measurement information or receive instructions, network devices for accessing network, monitoring devices between terminals and substation sites, and application servers which provides different application and so on devices.

   • Device failure: Field devices are distributed throughout the electric power system, such as SMs, intelligent electronic devices (IEDs), and RTUs. Despite the application of anti-virus software and firewalls, they may still be attacked by effective protocol messages, causing the channel's communication bandwidth to be exhausted, thereby rejecting incoming and outgoing data. Once the remaining memory space is insufficient, preventive measures can no longer continue to detect or prevent internal or external intrusions. Similarly, routers with limited computing power and storage space may be paralyzed due to data transmission delays, loss of sensitive data, service failures, and refusal to exchange data.

   • Unauthorized access to the device: The peripheral network devices of the power system, such as fax machines and connected modems, can be operated through unauthorized access control and dial-up access to field devices to achieve remote management. Since most devices do not require password verification or the default password is unchanged, an attacker can use this vulnerability to invade the network and further install a backdoor to access the prohibited area.

   • Work station interaction and fake human–machine interaction system disturbances: The SCADA network segment of the HMI system used by the SCADA system, redundant servers, workstations, and operators can unauthorized access to interrupt interaction with on-site equipment, or mislead operators to present another HMI interface for remote monitoring and control of remote equipment. The tampered software or middleware can control any device and network component, so that it can collect information of intelligent devices connected to the HMI system, delay data transmission in the CC, defect data exchange, and failure control operations, etc.

   • Damage on application services: Attackers can control application servers of different applications, send fault signals to system circuit breakers, isolate generators and shunts, and make the system perform load dumping operations. Meanwhile, due to the application server is damaged, the system operator may not be able to gain access to control commands. In addition, an attacker can illegally access the SCADA system through the communication links of unprotected remote network access points (e.g., backup facilities, development systems, and quality systems) to interfere with the corporate network.

2. **Data attacks:** An attacker maliciously inserts, changes, or deletes data or control commands in a network stream to mislead an operator into making a wrong decision or action. In general, the data that can be attacked includes the records of remote devices, the data transmitted by communication networks, or the control operations of applications.

   - Tampering with source data and system status: The source data is the data collected from the distributed remote terminal equipment and transmitted to the CC for analysis and decision-making. For example, SM power consumption reports, pricing signals, emergency messages, SM data, power outage reports, and fault notifications are transmitted to receiving devices through the communication network. The channel transmits information bidirectionally. During feed-forward and feedback, an attacker can inject fake information into the data stream transmitted in real-time in the channel to capture the output trace of the network system and delay the data transmission. In this case, the exchanged data may delay the response and mislead operations between field devices and other components, thereby reducing the overall performance of these devices. If the CC receives incorrect data from the field device, it may cause erroneous monitoring commands and operations, especially important operations such as sending a disconnect command, stopping power supply, or switching electrical equipment, thereby disrupting the balance between power demand and supply, resulting in the system shuts down abnormally, and the system runs beyond the limit, which may cause equipment failure or even casualties.
   - Destruction of CC data: The data of CC gathers control operation commands and application information. Due to limited communication bandwidth, redundant data streams may overwhelm the CC and block communication links. In addition, insufficient resources of the SCADA master station or slave station cause data transmission delay on the communication link, reduce the speed of the CC network, and interrupt the real-time service of the CC. For real-time operation, most measurement equipment (such as PMU) needs to be synchronized with the GPS. The attacker can forge the time stamp on the GPS signal, which can cause time error without accessing the PMU, which leads to the failure of the basic application of the PMU system, especially the phase angle error, which may lead to a series of undesirable system state variables. These state variables can be used to circumvent the BDD detection by coordinating the corresponding power flow measurement values, and passed to the CC for making decision-making commands. In addition,

the attacker may also deliberately tamper with the reported normal data to trigger the BDD mechanism, such as triggering a circuit breaker to cause erroneous operation, interrupting the normally running terminal equipment, and disrupting the system balance.

3. **Privacy attacks:** Aimed to learn or infer the user's personal information by analyzing the power usage data collected by SMs. SM stores logs and reports that summarize the user's power usage information, billing and pricing information, and reveals the operation behavior of SMs. Private information can be obtained in different ways. Attackers can exploit SMs's weak authentication mechanism to steal unencrypted messages and obtain passwords, thus preventing legitimate users from accessing the system normally, making it easier for attackers to process and answer requests from SM and utilities like valid users. At the same time, SM entertainment's information also discloses information about consumers' locations and other personal activities. In addition, non-customer power theft may cause non-technical losses by tampering with the use of time, pricing, physical events, and logs executing commands, as well as recorded network commands. However, changing user functionality can compromise user management modules that are configured for sensitive technologies such as communication lines and SMs data processing.

4. **Network reachability attacks:** Aimed to deplete or occupy limited communication and computing resources, interfere with data operations in communication networks, and cause delays or failures in data communications, a few seconds delay in data communication may invalidate control commands, cause equipment failure, and even cause an irreparable impact on the entire system. Network accessibility plays an important supporting role in the whole system. Communication network transmission is based on common standard messaging protocols, and attackers can use protocol analysis tools to sniff network traffic to collect unencrypted clear frames and then intercept SCADA DNP3 frames. Intercepted control and settings information is later used on another SCADA system or smart device, resulting in service interruption.

1.3.3   Research Motivations

The smart grid makes the ubiquitous power system and communication information network highly integrated. Through the Internet, Internet-of-Things, IED, and other information technologies, and with the help of cloud computing technology, the new economic model of Energy Internet sharing will be advanced.

The rapid and widespread incorporation of ICTs into the smart grid CPS, as mentioned in the Background section, has been introducing new vulnerabilities and threats even as we are taking actions to prevent, protect against, or minimize the impacts of known threats and hazards. At the same time, the growing dependence of the public, business, government, schools, hospitals, to name a few, on reliable and secure electricity has significantly increased the overall sensitivity to the impacts of any type of power instability. This, including voltage disturbances, momentary power outages, long-term service disruptions, and widespread blackouts with cascading effects, may, regardless of the causes, yield property damages, public health and safety dangers, and financial and life losses. Enhancing the security and resilience of the electric grid against malignant activities is critical to a functioning society.

Because of the complexity of the system and the dependence on information technology, the physical security of information has become an important bottleneck in the development of strong smart grid. High-speed two-way Communication Network is the foundation of Smart Grid, and Information Technology is the support. Only by fully mastering the data and information, can the resources be optimized and controlled reasonably, complete safe production and operation. The development of information physics fusion technology will lead to a new round of scientific and technological revolution and industrial transformation. In this monograph, we will mainly address the challenges of protecting data integrity to help build a secure smart grid CPS. Specifically, we will focus on analyzing the system reliability of a smart grid CPS under data integrity attacks as well as detection techniques of these data integrity attacks.

1.4   Research Contributions

The research focuses in this monograph lie in main topics relating to FDI attacks in smart grid CPSs including modeling and impact evaluation of FDI attacks, novel detection approaches for FDI attacks—both FmDI and FcDI attacks. Specifically, our main research contributions are summarized as follows:

- In Chap. 3, a stochastic Petri net based analytical model is developed to evaluate and analyze the system reliability of smart grid CPSs, specifically against topology attacks under system countermeasures (i.e., intrusion detection systems and malfunction recovery techniques). Topology attacks are evolved from FDI attacks, where attackers initialize FDI attacks by tempering with both measurement data and grid topology information. This analytical model is featured by bolstering both transient and steady-state analyses of system reliability.

- In Chap. 4, a distributed host-based collaborative detection scheme is proposed to detect FmDI attacks in smart grid CPSs. It is considered in this chapter that PMUs, deployed to measure the operating states of power grids, can be compromised by FmDI attackers, and the trusted host monitors (HMs) assigned to each PMU are employed to monitor and assess PMUs' behaviors. Neighboring HMs make use of the majority voting algorithm based on a set of predefined normal behavior rules to identify the existence of abnormal measurement data collected by PMUs. In addition, an innovative reputation system with an adaptive reputation updating algorithm is also designed to evaluate the overall operating status of PMUs, by which FmDI attacks as well as the attackers can be distinctly observed.

- In Chap. 5, a Dirichlet-based detection scheme for FcDI attacks in hierarchical smart grid CPSs is proposed. In the future hierarchical paradigm of a smart grid CPS, it is considered that the decentralized local agents (LAs) responsible for local management and control can be compromised by FcDI attackers. By issuing fake or biased commands, the attackers anticipate to manipulate the regional electricity prices with the purpose of illicit financial gains. The proposed scheme builds a Dirichlet-based probabilistic model to assess the reputation levels of LAs. This probabilistic model, used in conjunction with a designed adaptive reputation incentive mechanism, enables quick and efficient detection of FcDI attacks as well as the attackers.
- In Chap. 6, we systematically explore the feasibility and limitations of detecting FmDI attacks in smart grid CPSs using distributed flexible AC transmission system (D-FACTS) devices. Recent studies have investigated the possibilities of proactively detecting FmDI attacks on smart grid CPSs by using D-FACTS devices. We term this approach as proactive false data detection (PFDD). In this chapter, the feasibility of using PFDD to detect FmDI attacks is investigated by considering single-bus, uncoordinated multiple-bus, and coordinated multiple-bus FmDI attacks, respectively. It is proved that PFDD can detect all these three types of FmDI attacks targeted on buses or super-buses with degrees larger than 1, as long as the deployment of D-FACTS devices covers branches at least containing a spanning tree of the grid graph. The minimum efforts required for activating D-FACTS devices to detect each type of FmDI attacks are, respectively, evaluated. In addition, the limitations of this approach are also discussed, and it is strictly proved that PFDD is not able to detect FmDI attacks targeted on buses or super-buses with degrees equalling 1.

1.5 Monograph Outline

The organization of the remainder of this monograph is as follows. Chapter 2 introduces some fundamentals relating to FDI attacks in smart grid CPSs, and provides the state-of-the-art literature reviews. Chapter 3 models the attack strategy of topology attacks using a stochastic Petri net approach and analyzes the system reliability under such attacks. Chapter 4 presents a novel distributed host-based collaborative detection scheme for FmDI attacks. In Chap. 5, a Dirichlet-based detector for FcDI attacks as well as the compromised insiders is introduced, followed by the discussion of feasibility and limitations of detecting FmDI attacks using D-FACTS devices in Chap. 6. Chapter 7 concludes the monograph and briefs some promising research directions for future work.

References

1. Moslehi, K., & Kumar, R. (June 2010). A reliability perspective of the smart grid. *IEEE Transactions on Smart Grid, 1*(1), 57–64.
2. Farhangi, H. (Jan.-Feb. 2010). The path of the smart grid. *IEEE Power and Energy Magazine, 8*(1), 18–28.
3. Amin, S. M., & Wollenberg, B. F. (Sep.-Oct. 2005). Toward a smart grid: power delivery for the 21st century. *IEEE Power and Energy Magazine, 3*(5), 34–41.
4. Li, X., Liang, X., Lu, R., Shen, X., Lin, X., & Zhu, H. (Aug. 2012). Securing smart grid: cyber attacks, countermeasures, and challenges. *IEEE Communications Magazine, 50*(8), 38–45.
5. Farwell, J. P., & Rohozinski, R. (Feb. 2011). Stuxnet and the future of cyber war. *Survival, 53*(1), 23–40.
6. Falliere, N., Murchu, L. O., & Chien, E. (Feb. 2011). W32. Stuxnet dossier. *White Paper, Symantec Corp., Security Response, 5*(6), 29.
7. Cloherty, J., & Thomas, P. (Nov. 2014). 'Trojan Horse' bug lurking in vital US computers since 2011. *ABC News*. [Online]. Available: https://abcnews.go.com/US/trojan-horse-bug-lurking-vital-us-computers-2011/story?id=26737476
8. Perlroth, N. (Oct. 2012). In cyberattack on Saudi firm, U.S. sees Iran firing back. *The New York Times*. [Online]. Available: https://www.nytimes.com/2012/10/24/business/global/cyberattack-on-saudi-oil-firm-disquiets-us.html
9. Goldman, J. (Feb. 2013). In cyberattack on Saudi firm, U.S. sees Iran firing back. *eSecurity Planet*. [Online]. Available: https://www.esecurityplanet.com/network-security/florida-utility-company-hit-by-cyber-attack.html
10. Zetter, K. (Mar. 2016). Inside the cunning, unprecedented hack of Ukraine's power grid. *Wired*. [Online]. Available: https://www.wired.com/2016/03/inside-cunning-unprecedented-hack-ukraines-power-grid/
11. Johnson, B., Caban, D., Krotofil, M., Scali, D., Brubaker, N., & Glyer, C. (Dec. 2017). Attackers deploy new ICS attack framework "TRITON" and cause operational disruption to critical infrastructure. *FireEye*. [Online]. Available: https://www.fireeye.com/blog/threat-research/2017/12/attackers-deploy-new-ics-attack-framework-triton.html
12. U.S.. (Dec. 2016). National electric grid security and resilience action plan. *Executive Office of the President, U.S.*. [Online]. Available: https://www.whitehouse.gov/sites/whitehouse.gov/files/images/National_Electric_Grid_Action_Plan_06Dec2016.pdf
13. Canada. (Dec. 2016). National electric grid security and resilience action plan. *Government of Canada*. [Online]. Available: https://www.whitehouse.gov/sites/whitehouse.gov/files/images/National_Electric_Grid_Action_Plan_06Dec2016.pdf
14. Reuters. (Sep. 2015). UPDATE 1-china targets $300 bln power grid spend over 2015-20 - report. *Reuters*. [Online]. Available: https://www.reuters.com/article/china-power-transmission-idUSL4N1171UP20150901
15. Australia. (2018). A better energy future for Australia. *Department of the Environment and Energy, Australia*. [Online]. Available: https://www.energy.gov.au/government-priorities/better-energy-future-australia
16. UK. (Nov. 2012). Energy security strategy. *Department of Energy and Climate Change, UK*. [Online]. Available: https://assets.publishing.service.gov.uk/government/uploads/system/uploads/attachment$_$data/file/65643/7101-energy-security-strategy.pdf
17. Hamilton, B. A., Miller, J., & Renz, B. (June 2010). Understanding the benefits of the smart grid-smart grid implementation strategy. *United States: United States Department of Energy's National Energy Technology Laboratory*. [Online]. Available: https://www.netl.doe.gov/File%20Library/research/energy%20efficiency/smart%20grid/whitepapers/06-18-2010$_$Understanding-Smart-Grid-Benefits.pdf
18. Li, W. (Mar. 2014). *Risk assessment of power systems: models, methods, and applications*. New York: John Wiley & Sons.
19. Lu, R. (May 2016). *Privacy-enhancing aggregation techniques for smart grid communications*. New York: Springer.

Chapter 2
Fundamentals and Related Literature

In this chapter, we will introduce some fundamental concepts serving as the building blocks of our research work, which include the state estimation, bad data detection, as well as FmDI attacks against state estimation. In addition, the state-of-the-art literature reviews on existing intrusion detection systems (IDSs), FDI attacks detection, and insider threats detection will also be provided.

2.1 State Estimation and Bad Data Detection

In this section, we introduce the fundamental concepts of state estimation and bad data detection, one of the most important techniques in electric grids.

2.1.1 AC Power Flow Model State Estimation Formulation

In a power system, state estimation is a process that estimates the current system state through analysis of the measurements collected from a large number of meters and power system models, it is the heart of control systems to support real-time analysis, contingency analysis, and power management and control. Specifically, the operator in the control center will perform contingency analysis, with the help of the state estimation output, including inferring potential operational problems, the action could avoid those problems, and the potential risks of those actions. Power flow models, which are constructed by a set of equations that depict the energy flow, are used for state estimation. The basic relationship, with an AC power flow model, between the measurement data and system states is given by Schweppe and Rom [1]

© Springer Nature Switzerland AG 2020
B. Li et al., *Detection of False Data Injection Attacks in Smart Grid Cyber-Physical Systems*, Wireless Networks,
https://doi.org/10.1007/978-3-030-58672-0_2

$$\mathbf{z} = \mathbf{h}(\mathbf{x}) + \eta, \tag{2.1}$$

where $\mathbf{z} \in \mathbb{R}^{m \times 1}$ is the measurement vector containing information of power generations, power loads, and power flows, $\mathbf{x} \in \mathbb{R}^{n \times 1}$ is the system state vector including bus voltage phase angles, and $\eta \in \mathbb{R}^{m \times 1}$ is the measurement noise vector with zero mean and covariance $\mathbf{W} \in \mathbb{R}^{m \times m}$, a diagonal matrix including reciprocals of the variances of meter errors, which is given by

$$\mathbf{W} = \begin{bmatrix} \sigma_1^{-2} & & & \\ & \sigma_2^{-2} & & \\ & & \cdot & \\ & & & \cdot & \\ & & & & \sigma_m^{-2} \end{bmatrix}, \tag{2.2}$$

where σ_i^{-2} is the i-th meter variance ($1 \leq i \leq m$). Note that m and n are the numbers of measurements and system states, respectively, and $m > n$ indicates that redundant measurements are introduced. $\mathbf{h}(\mathbf{x}) = (h_1(\mathbf{x}), \ldots, h_m(\mathbf{x}))^{\mathsf{T}}$ and $h_i(\mathbf{x})$ is a nonlinear function of $\mathbf{x}$, which relates the system states to the ideal measurements. The object of the state estimation is to find the estimated system state vector $\hat{\mathbf{x}}$. Based on the AC power flow model, the $\hat{\mathbf{x}}$ can be found through minimizing the least squares which is given by

$$\hat{\mathbf{x}} = \arg \min_{\mathbf{x}} (\mathbf{z} - \mathbf{h}(\mathbf{x}))^{\mathsf{T}} \mathbf{W}^{-1} (\mathbf{z} - \mathbf{h}(\mathbf{x})). \tag{2.3}$$

The AC state estimation is, in reality, highly nonlinear and usually implemented iteratively. The Newton–Raphson iteration, for example, can be used until the solution converges. However, this process is time consuming and cannot guarantee a convergence to the global optimal value.

2.1.2 DC Power Flow Model State Estimation Formulation

Since state estimation is usually applied over the high-voltage power transmission networks, it is reasonable to approximate AC power flow model to a DC one [2]. In this way, the measurement data and system states are related by

$$\mathbf{z} = \mathbf{H}\mathbf{x} + \eta, \tag{2.4}$$

where $\mathbf{H} \in \mathbb{R}^{m \times n}$ is the measurement Jacobian matrix, implying the system connection and configuration information. Our research studies in later chapters are all based on DC state estimation. Although AC power flow model is more accurate than DC model, it is computationally expensive and too complex to be used in the analysis. In contrast, DC power flow model is much faster, more robust, and techno-economic than AC, and it has been widely accepted as a useful simplification of AC model [2–7].

Based on the DC power flow model, with the relationship shown in Eq. (2.4), the estimated system state vector $\hat{\mathbf{x}}$ using the least squares is given by

$$\hat{\mathbf{x}} = \arg\min_{\mathbf{x}}(\mathbf{z} - \mathbf{H}\mathbf{x})^{\mathsf{T}}\mathbf{W}^{-1}(\mathbf{z} - \mathbf{H}\mathbf{x}). \tag{2.5}$$

The linear DC state estimation has a closed-form solution obtained through a non-iterative procedure by solving Eq. (2.5), which is given by

$$\hat{\mathbf{x}} = (\mathbf{H}^{\mathsf{T}}\mathbf{W}^{-1}\mathbf{H})^{-1}\mathbf{H}^{\mathsf{T}}\mathbf{W}^{-1}\mathbf{z} \triangleq \mathbf{\Lambda}\mathbf{z}, \tag{2.6}$$

where

$$\mathbf{\Lambda} \triangleq (\mathbf{H}^{\mathsf{T}}\mathbf{W}^{-1}\mathbf{H})^{-1}\mathbf{H}^{\mathsf{T}}\mathbf{W}^{-1} \tag{2.7}$$

is the DC state estimator as well as the "pseudo-inverse" of $\mathbf{H}$. Besides the above-mentioned criterion, some other statistical estimation criteria, such as the maximum likelihood criterion and the minimum variance criterion, are also commonly used in the DC state estimation. These criteria will result in the identical optimal state estimator $\mathbf{\Lambda}$, if measurement errors are assumed to follow the normal distribution with zero mean [8]. Since the unique $\hat{\mathbf{x}}$ can be obtained when $\mathbf{H}$ is a full rank matrix, at least n meter measurements are required to gain a unique $\hat{\mathbf{x}}$. Therefore, the n needed meter measurements are regarded as basic meter measurements, and the other $(m - n)$ meter measurements are utilized to tackle the random measurement noises received by the control center. Then the estimated measurement data $\hat{\mathbf{z}}$ is given by

$$\hat{\mathbf{z}} = \mathbf{H}\hat{\mathbf{x}} = \mathbf{H}\mathbf{\Lambda}\mathbf{z}. \tag{2.8}$$

2.1.3   General Structure of H matrix

The measurement Jacobian $\mathbf{H}$ is not a square matrix. In general, for full measurement set, the Jacobian $\mathbf{H}$ is a matrix with $(2N - 1)$ columns and $(3N + 4B)$ rows [9], where N and B are numbers of buses and lines, respectively. To construct the $\mathbf{H}$ matrix, initially all its elements are set to zero. Network elements are considered one-by-one.

As shown in Fig. 2.1, the admittance of the series branch connecting buses i and j is described by the admittance $g_{ij} + jb_{ij}$, the external shunt admittance at buses

Fig. 2.1 Two-port π-model
of a network branch [10]

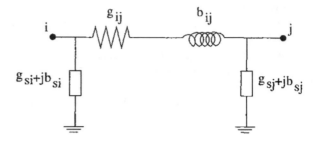

i and j is described as $g_{si} + jb_{si}$ and $g_{sj} + jb_{sj}$, respectively. For such a model of
the network branch, the real and reactive power flow in line i-j are given by

$$p_{ij} = V_i^2(g_{si} + g_{ij}) - V_i V_j(g_{ij} \cos\theta_{ij} + b_{ij} \sin\theta_{ij}), \tag{2.9}$$

$$p_{ji} = V_j^2(g_{sj} + g_{ij}) - V_i V_j(g_{ij} \cos\theta_{ij} - b_{ij} \sin\theta_{ij}), \tag{2.10}$$

$$q_{ij} = -V_i^2(b_{si} + b_{ij}) - V_i V_j(g_{ij} \sin\theta_{ij} - b_{ij} \cos\theta_{ij}), \tag{2.11}$$

$$q_{ji} = -V_j^2(b_{sj} + b_{ij}) + V_i V_j(g_{ij} \sin\theta_{ij} + b_{ij} \cos\theta_{ij}), \tag{2.12}$$

where V_i, θ_i represent the bus i voltage magnitude and phase angle, and $\theta_{ij} = \theta_i - \theta_j$. Then, the real and reactive power flows on all of the lines, which are computed
from Eqs. (2.9)–(2.12), are stored in matrix as shown in Eqs. (2.13) and (2.14) by
which bus powers can be given.

$$\mathbf{P} = \begin{bmatrix} 0 & p_{12} & p_{13} & \cdots & p_{1N} \\ p_{21} & 0 & p_{23} & \cdots & p_{2N} \\ p_{31} & p_{32} & 0 & \cdots & p_{3N} \\ \vdots & \vdots & \vdots & & \vdots \\ p_{N1} & p_{N2} & p_{N3} & \cdots & 0 \end{bmatrix}, \tag{2.13}$$

$$\mathbf{Q} = \begin{bmatrix} 0 & q_{12} & q_{13} & \cdots & q_{1N} \\ q_{21} & 0 & q_{23} & \cdots & q_{2N} \\ q_{31} & q_{32} & 0 & \cdots & q_{3N} \\ \vdots & \vdots & \vdots & & \vdots \\ q_{N1} & q_{N2} & q_{N3} & \cdots & 0 \end{bmatrix}. \tag{2.14}$$

After that, the general Jacobian $\mathbf{H}$ can be constructed by

$$
\mathbf{H} = \begin{bmatrix} \mathbf{H}_{V,\theta} & \mathbf{H}_{V,V} \\ \mathbf{H}_{p_{ij},\theta} & \mathbf{H}_{p_{ij},V} \\ \mathbf{H}_{p_{ji},\theta} & \mathbf{H}_{p_{ji},V} \\ \mathbf{H}_{q_{ij},\theta} & \mathbf{H}_{q_{ij},V} \\ \mathbf{H}_{q_{ji},\theta} & \mathbf{H}_{q_{ji},V} \\ \mathbf{H}_{P,\theta} & \mathbf{H}_{P,V} \\ \mathbf{H}_{Q,\theta} & \mathbf{H}_{Q,V} \end{bmatrix}.
\tag{2.15}
$$

In Jacobian $\mathbf{H}$, all the elements can be derived from Eqs. (2.9)–(2.12), and represent the partial derivatives of bus voltage magnitudes, bus powers and line flows with respect to state variables θ and V. The sub-matrices $\mathbf{H}_{V,\theta}$ and $\mathbf{H}_{V,V}$ are given by

$$
\mathbf{H}_{V,\theta} = \begin{bmatrix} \frac{\partial |V_1|}{\partial \theta_2} & \frac{\partial |V_1|}{\partial \theta_3} & \cdots & \frac{\partial |V_1|}{\partial \theta_N} \\ \frac{\partial |V_2|}{\partial \theta_2} & \frac{\partial |V_2|}{\partial \theta_3} & \cdots & \frac{\partial |V_2|}{\partial \theta_N} \\ \vdots & \vdots & \vdots & \vdots \\ \frac{\partial |V_N|}{\partial \theta_2} & \frac{\partial |V_N|}{\partial \theta_3} & \cdots & \frac{\partial |V_N|}{\partial \theta_N} \end{bmatrix},
\tag{2.16}
$$

$$
\mathbf{H}_{V,V} = \begin{bmatrix} \frac{\partial |V_1|}{\partial |V_1|} & \frac{\partial |V_1|}{\partial |V_2|} & \cdots & \frac{\partial |V_1|}{\partial |V_N|} \\ \frac{\partial |V_2|}{\partial |V_1|} & \frac{\partial |V_2|}{\partial |V_2|} & \cdots & \frac{\partial |V_2|}{\partial |V_N|} \\ \vdots & \vdots & \vdots & \vdots \\ \frac{\partial |V_N|}{\partial |V_1|} & \frac{\partial |V_N|}{\partial |V_2|} & \cdots & \frac{\partial |V_N|}{\partial |V_N|} \end{bmatrix},
\tag{2.17}
$$

where the $\mathbf{H}_{V,\theta}$ is a zero matrix, since $\frac{\partial |V_i|}{\partial \theta_j} = 0$ for all i and j, and $\frac{\partial |V_i|}{\partial |V_j|}$ in the matrix $\mathbf{H}_{V,V}$ is given by

$$
\frac{\partial V_i}{\partial V_j} = \begin{cases} 0, & \text{if } i \neq j, \\ 1, & \text{otherwise.} \end{cases}
\tag{2.18}
$$

Note that the row about a particular bus will be deleted if the voltage meter at the particular bus is not available. The sub-matrices $\mathbf{H}_{p_{ij},\theta}$ and $\mathbf{H}_{p_{ij},V}$ corresponding to real power flow measurements are given by

$$
\mathbf{H}_{p_{ij},\theta} = \begin{bmatrix} & \vdots & & \vdots & \\ \cdots & \frac{\partial p_{ij}}{\partial \theta_i} & \cdots & \frac{\partial p_{ij}}{\partial \theta_j} & \cdots \\ & \vdots & & \vdots & \end{bmatrix},
\tag{2.19}
$$

$$\mathbf{H}_{p_{ij},V} = \begin{bmatrix} & \vdots & & \vdots & \\ \cdots & \frac{\partial p_{ij}}{\partial |V_i|} & \cdots & \frac{\partial p_{ij}}{\partial |V_j|} & \cdots \\ & \vdots & & \vdots & \end{bmatrix}, \tag{2.20}$$

where the partial derivatives of p_{ij} corresponding to θ_i, θ_j, V_i, and V_j can be obtained by using Eq. (2.9). Thus, we have

$$\frac{\partial p_{ij}}{\partial \theta_i} = V_i V_j (g_{ij} \sin \theta_{ij} - b_{ij} \cos \theta_{ij}), \tag{2.21}$$

$$\frac{\partial p_{ij}}{\partial \theta_j} = -V_i V_j (g_{ij} \sin \theta_{ij} - b_{ij} \cos \theta_{ij}), \tag{2.22}$$

$$\frac{\partial p_{ij}}{\partial V_i} = -V_j (g_{ij} \cos \theta_{ij} + b_{ij} \sin \theta_{ij}) + 2V_i (g_{ij} + g_{si}), \tag{2.23}$$

$$\frac{\partial p_{ij}}{\partial V_j} = -V_i (g_{ij} \cos \theta_{ij} + b_{ij} \sin \theta_{ij}). \tag{2.24}$$

The sub-matrices $\mathbf{H}_{q_{ij},\theta}$ and $\mathbf{H}_{q_{ij},V}$ corresponding to reactive power flow measurements are given by

$$\mathbf{H}_{q_{ij},\theta} = \begin{bmatrix} & \vdots & & \vdots & \\ \cdots & \frac{\partial q_{ij}}{\partial \theta_i} & \cdots & \frac{\partial q_{ij}}{\partial \theta_j} & \cdots \\ & \vdots & & \vdots & \end{bmatrix}, \tag{2.25}$$

$$\mathbf{H}_{q_{ij},V} = \begin{bmatrix} & \vdots & & \vdots & \\ \cdots & \frac{\partial q_{ij}}{\partial |V_i|} & \cdots & \frac{\partial q_{ij}}{\partial |V_j|} & \cdots \\ & \vdots & & \vdots & \end{bmatrix}, \tag{2.26}$$

where the partial derivatives of q_{ij} corresponding to θ_i, θ_j, V_i, and V_j can be obtained by using Eq. (2.9). Thus

$$\frac{\partial q_{ij}}{\partial \theta_i} = -V_i V_j (g_{ij} \cos \theta_{ij} + b_{ij} \sin \theta_{ij}), \tag{2.27}$$

$$\frac{\partial q_{ij}}{\partial \theta_j} = V_i V_j (g_{ij} \cos \theta_{ij} + b_{ij} \sin \theta_{ij}), \tag{2.28}$$

$$\frac{\partial q_{ij}}{\partial V_i} = -V_j (g_{ij} \sin \theta_{ij} - b_{ij} \cos \theta_{ij}) - 2V_i (b_{ij} + b_{si}), \tag{2.29}$$

$$\frac{\partial q_{ij}}{\partial V_j} = -V_i (g_{ij} \sin \theta_{ij} - b_{i}j \cos \theta_{ij}). \tag{2.30}$$

The sub-matrices $\mathbf{H}_{p_{ji},\theta}$, $\mathbf{H}_{p_{ji},V}$, $\mathbf{H}_{q_{ji},\theta}$, and $\mathbf{H}_{q_{ji},V}$ can be constructed through similar methods. The sub-matrices $\mathbf{H}_{P,\theta}$, $\mathbf{H}_{P,V}$, $\mathbf{H}_{Q,\theta}$, and $\mathbf{H}_{Q,V}$ containing partial derivatives of bus powers can be constructed by partial of derivatives of line flows. For instance, the sub-matrices $\mathbf{H}_{P,\theta}$ and $\mathbf{H}_{Q,\theta}$ can be given by

$$\mathbf{H}_{P,\theta} = \begin{bmatrix} \frac{\partial P_1}{\partial \theta_2} & \frac{\partial P_1}{\partial \theta_3} & \cdots & \frac{\partial P_1}{\partial \theta_N} \\ \frac{\partial P_2}{\partial \theta_2} & \frac{\partial P_2}{\partial \theta_3} & \cdots & \frac{\partial P_2}{\partial \theta_N} \\ \vdots & \vdots & \vdots & \vdots \\ \frac{\partial P_N}{\partial \theta_2} & \frac{\partial P_N}{\partial \theta_3} & \cdots & \frac{\partial P_N}{\partial \theta_N} \end{bmatrix}, \tag{2.31}$$

$$\mathbf{H}_{Q,\theta} = \begin{bmatrix} \frac{\partial Q_1}{\partial \theta_2} & \frac{\partial Q_1}{\partial \theta_3} & \cdots & \frac{\partial Q_1}{\partial \theta_N} \\ \frac{\partial Q_2}{\partial \theta_2} & \frac{\partial Q_2}{\partial \theta_3} & \cdots & \frac{\partial Q_2}{\partial \theta_N} \\ \vdots & \vdots & \vdots & \vdots \\ \frac{\partial Q_N}{\partial \theta_2} & \frac{\partial Q_N}{\partial \theta_3} & \cdots & \frac{\partial Q_N}{\partial \theta_N} \end{bmatrix}, \tag{2.32}$$

where P_i and Q_i are the real and reactive power injection measurements at bus i, respectively. To illustrate this, the elements connected at bus i are set as j, k, and m. In such a case, the P_i and Q_i can be calculated by

$$P_i = p_{ij} + p_{ik} + p_{im}, \tag{2.33}$$

$$Q_i = q_{ij} + q_{ik} + q_{im}. \tag{2.34}$$

Then, $\frac{\partial P_i}{\partial \theta_i}$ and $\frac{\partial Q_i}{\partial \theta_i}$ can be given by

$$\frac{\partial P_i}{\partial \theta_i} = \frac{\partial p_{ij}}{\partial \theta_i} + \frac{\partial p_{ik}}{\partial \theta_i} + \frac{\partial p_{im}}{\partial \theta_i}, \tag{2.35}$$

$$\frac{\partial Q_i}{\partial \theta_i} = \frac{\partial q_{ij}}{\partial \theta_i} + \frac{\partial q_{ik}}{\partial \theta_i} + \frac{\partial q_{im}}{\partial \theta_i}. \tag{2.36}$$

Similar to $\mathbf{H}_{P,\theta}$ and $\mathbf{H}_{Q,\theta}$, the sub-matrices $\mathbf{H}_{P,V}$ and $\mathbf{H}_{Q,V}$ can also be obtained by partial of derivatives of line flows. Once all the sub-matrices are obtained, a general Jacobian $\mathbf{H}$ can be constructed completely.

In practice, since not all measurements are available, the number of rows in the Jacobian $\mathbf{H}$ will be much less than $(3N + 4B)$. It is, therefore, computing and recording only required elements, other than all elements are usually preferred. In the case of the two-port model, all the partial derivatives in sub-matrices of **H** are not always necessary to be completely calculated. For example, the partial derivatives $\frac{\partial p_{ij}}{\partial \theta_i}$, $\frac{\partial p_{ij}}{\partial \theta_j}$, $\frac{\partial p_{ij}}{\partial V_i}$, and $\frac{\partial p_{ij}}{V_j}$ will be computed only if p_{ij} or P_i or both p_{ij} and P_i exist in the measurement list. Therefore, there are three cases:

- p_{ij} is available: the four partial derivatives mentioned above are added to the row corresponding to p_{ij}.
- P_i is available: the four partial derivatives mentioned above are taken as the previous values of the row corresponding to P_i.
- Both p_{ij} and P_i are available: the four partial derivatives mentioned above are added to the row corresponding to p_{ij} and taken as the previous values of the row corresponding to P_i.

Likewise, $\frac{\partial p_{ji}}{\partial \theta_i}$, $\frac{\partial p_{ji}}{\partial \theta_j}$, $\frac{\partial p_{ji}}{\partial V_i}$, $\frac{\partial p_{ji}}{\partial V_j}$, $\frac{\partial q_{ij}}{\partial \theta_i}$, $\frac{\partial q_{ij}}{\partial \theta_j}$, $\frac{\partial q_{ij}}{\partial V_i}$, $\frac{\partial q_{ij}}{\partial V_j}$ and $\frac{\partial q_{ji}}{\partial \theta_i}$, $\frac{\partial q_{ji}}{\partial \theta_j}$, $\frac{\partial q_{ji}}{\partial V_i}$, $\frac{\partial q_{ji}}{\partial V_i}$, $\frac{\partial q_{ji}}{\partial V_j}$ can be computed in such ways [10].

2.1.4   Bad Data Detection

The existing bad data detection approaches usually use the hypothesis testing, by observing the largest normalized residual (LNR) to detect the bad measurement data. The normalized measurement residual $\gamma \in \mathbb{R}^{m \times 1}$ is calculated based on the difference between the measurement data $\mathbf{z}$ and the estimated measurement data $\hat{\mathbf{z}}$, i.e.,

$$\gamma = (z - \hat{z}) = (z - H \Lambda z) = (I - H\Lambda)z, \tag{2.37}$$

where $\mathbf{I} \in \mathbb{R}^{m \times m}$ is the identity matrix. The hypothesis testing is expressed as

$$\begin{cases} \text{Null hypothesis } \mathbf{H}_0 : \|\overline{\gamma}\| > \tau \\ \text{Alternative hypothesis } \mathbf{H}_1 : \|\overline{\gamma}\| <= \tau, \end{cases} \tag{2.38}$$

where $\overline{\gamma} = \sqrt{\mathbf{W}^{-1}}\gamma$ is the normalized measurement residual vector [8]. This testing is to compare the Frobenius norm of the normalized measurement residual $\|\overline{\gamma}\|$ with a predefined threshold τ. Specifically, if $\|\overline{\gamma}\| > \tau$, the null hypothesis is accepted indicating the existence of anomalous residuals; hence, bad measurement data is present in $\mathbf{z}$. Otherwise (i.e., $\|\overline{\gamma}\| <= \tau$), the null hypothesis is rejected, which implies no bad measurement data. The value of τ can be determined by a chi-squared test with a significance level of α, i.e., $\tau = \sqrt{\chi^2_{m-n,1-\alpha}}$ because

$\|\overline{\gamma}\|^2 = \|\sqrt{W^{-1}}\gamma\|^2 = \|\sqrt{W^{-1}}(z - H\hat{x})\|^2$ follows a chi-square distribution $\mathcal{X}^2_{m-n}$, where $m - n$ is the degree of freedom [8].

2.2 FmDI Attacks Against AC State Estimation

In 2011, Liu et al. demonstrated that a set of smart attackers can initiate FmDI attacks in electric grids against the existing state estimation and bad data detection technique, as long as they can compromise some meter devices and have some knowledge of electric grid connections and configurations [11]. The authors consider two attack scenarios, including

- Limited access to meters: in this attack scenario, the attacker can only compromise certain meters due to stronger physical protection of the meters, e.g., considering the difficulty of intrusion, attackers are more likely to access meters located outside the building rather than those located in substations with physical perimeter control.
- Limited resource available to compromise meters: in this attack scenario, the attacker can only utilize the limited resource to compromise meters, e.g., the resource owned by the attacker can only compromise up to k meters; therefore, the attacker may also want to initiate FmDI attacks with minimizing the number of meters.

The goal of FmDI attacks, which is different from cyberattacks (e.g., flooding, jamming, eavesdropping, and denial-of-service attacks) that destroy data availability or confidentiality, is to destroy data integrity. Besides, FmDI attacks can bypass the existing safeguards, that is to say, they are different from other types of cyberattacks that are likely to be detected by traditional detection methods. The bad data detection is effective when the malicious data a is unstructured. However, in the case of FmDI attacks, since the attack vector cannot be distinguished effectively from the original estimated system state vector, these threats are "unobservable" to the control center.

In the case of AC power flow model, to successfully initiate FmDI attacks, the attacker needs to influent at least one state variable. That is, all the measurements corresponding to this state variable need to be manipulated. In DC state estimation, the relationship between state variables and measurements is linear, i.e., a nonzero element ij of matrix H indicates that the function z_i directly corresponds to state x_j. In AC state estimation, as shown in Eq. (2.1), such a relationship is nonlinear. However, the information that indicates which measurement is directly related to which state variable can be provided by the Jacobian matrix of $h(x)$, i.e.,

$$
\mathbf{J}_h = \begin{bmatrix}
\frac{\partial h_1(\mathbf{x})}{\partial x_1} & \frac{\partial h_1(\mathbf{x})}{\partial x_2} & \cdots & \frac{\partial h_1(\mathbf{x})}{\partial x_n} \\[2mm]
\frac{\partial h_2(\mathbf{x})}{\partial x_1} & \frac{\partial h_2(\mathbf{x})}{\partial x_2} & \cdots & \frac{\partial h_2(\mathbf{x})}{\partial x_n} \\[2mm]
\vdots & \vdots & \vdots & \vdots \\[2mm]
\frac{\partial h_m(\mathbf{x})}{\partial x_1} & \frac{\partial h_m(\mathbf{x})}{\partial x_2} & \cdots & \frac{\partial h_m(\mathbf{x})}{\partial x_n}
\end{bmatrix}, \tag{2.39}
$$

where $h_i(\mathbf{x})$ and x_i are the i-th element of $\mathbf{h}(\mathbf{x})$ and $\mathbf{x}$, respectively. The element of $\mathbf{J}_h$ which is in the row corresponding to the measurement and in the column corresponding to the state variable must be nonzero if the measurement is dependent on the state variable. Otherwise, the element is equal to zero. Thus, the assaulter can determine the minimum number of measurements that need to be tampered with through considering the row corresponding to the targeted measurement and columns with nonzero elements in the row, i.e., such a minimum number of measurements depend on the minimum number of nonzero elements present in any of these columns. Note that there is a premise that all the buses have power injection.

After determining the measurements have to be tampered with, the attacker will determine what values they need to be modified to. In the case of the series branch connecting buses i and j (see Fig. 2.1), suppose the attacker wants to choose V_i as the target state variable need to be tampered with, then the real power flow p'_{ij} shown in the following equation will yield the influenced voltage magnitude:

$$
p'_{ij} = V_{i,a}^2(g_{si}) + g_{ij}) - V_{i,a}V_j(g_{ij}\cos\theta_{ij} + b_{ij}\sin\theta_{ij}), \tag{2.40}
$$

where $V_{i,a}$ is the state variable influenced by the attacker. There are similar logics for other real and reactive power flows. The quadratic Eq. (2.40) has multiple solutions, while only one of them has physical meaning. To solve the above equation, the value $V_{i,a}$, V_j, θ_{ij} need to be determined or estimated by calculating p_{ij}, q_{ij}, P_i, and Q_i for the other measurements with methods mentioned in Sect. 2.1.

Another assumption is that the attacker wants to choose θ_i as the target state variable needs to be tampered with. Since the active power flow and injection are more sensitive to changes in voltage angles than in voltage magnitudes, smaller voltage angle changes will have a greater impact on power flows. There will be more possible types of solutions if attackers choose both the voltage magnitude and angle as targets to be tampered with.

To construct an FmDI attack, the attacker needs to design an attack vector $\mathbf{a} \in \mathbb{R}^{m \times 1}$ and fabricate a malicious measurement vector $\mathbf{z_a} = \mathbf{z} + \mathbf{a}$. The Frobenius norm of the normalized measurement residual with false data injected $\|\overline{\boldsymbol{\gamma}}_a\|$ is given by

$$\|\overline{\pmb{\gamma}}_a\| = \|\sqrt{\mathbf{W}^{-1}}[\mathbf{z_a} - \mathbf{h}(\hat{\mathbf{x}}_\mathbf{a})]\|$$

$$= \|\sqrt{\mathbf{W}^{-1}}[\mathbf{z} + \mathbf{a} - \mathbf{h}(\hat{\mathbf{x}} + \mathbf{c})]\|$$

$$= \left\|\sqrt{\mathbf{W}^{-1}}\left[\begin{pmatrix} \mathbf{z_o} \\ \mathbf{z_t} + \mathbf{a_t} \end{pmatrix} - \begin{pmatrix} \mathbf{h_o}(\hat{\mathbf{x}}_\mathbf{o}) \\ \mathbf{h_t}(\hat{\mathbf{x}}_\mathbf{o}, \hat{\mathbf{x}}_\mathbf{t} + \mathbf{c}) \end{pmatrix}\right]\right\| \qquad (2.41)$$

$$= \left\|\sqrt{\mathbf{W}^{-1}}\left[\begin{pmatrix} \mathbf{z_o} \\ \mathbf{z_t} \end{pmatrix} - \begin{pmatrix} \mathbf{h_o}(\hat{\mathbf{x}}_\mathbf{o}) \\ \mathbf{h_t}(\hat{\mathbf{x}}_\mathbf{o}, \hat{\mathbf{x}}_\mathbf{t}) \end{pmatrix}\right]\right\|$$

$$= \|\sqrt{\mathbf{W}^{-1}}[\mathbf{z} - \mathbf{h}(\hat{\mathbf{x}})]\| \leq \tau,$$

where variables with the subscript o indicate the measurements or state variables that are not impacted by the attacker, variables with the subscript t indicate those that were maliciously impacted, the vector $\mathbf{c} \in \mathbb{R}^{n \times 1}$ is the required changes of the estimated state variables. Besides, in order to ensure that the attack can successfully bypass the bad data detection, the following equation needs to be satisfied:

$$\mathbf{a_t} = \mathbf{h_t}(\hat{\mathbf{x}}_\mathbf{o}, \hat{\mathbf{x}}_\mathbf{t} + \mathbf{c}) - \mathbf{h_t}(\hat{\mathbf{x}}_\mathbf{o}, \hat{\mathbf{x}}_\mathbf{t}). \qquad (2.42)$$

As shown in the equation above, the attacker must know a set of estimated state variables from $\mathbf{h_t}$.

2.3 FmDI Attacks Against DC State Estimation

Similar to the FmDI attacks on AC state estimation, if there exists a vector $\mathbf{c} \in \mathbb{R}^{n \times 1}$ that can satisfy $\mathbf{a} = \mathbf{Hc}$, a successful FmDI is constructed and the original estimated system state vector $\hat{\mathbf{x}}$ is injected with an offset by $\hat{\mathbf{x}}_\mathbf{a} = \hat{\mathbf{x}} + \mathbf{c}$. This is because that with such false data being injected, the estimated system states vector $\hat{\mathbf{x}}_\mathbf{a}$ with reference to Eq. (2.6) is given by

$$\hat{\mathbf{x}}_\mathbf{a} = \mathbf{\Lambda}\mathbf{z_a} = \mathbf{\Lambda}(\mathbf{z} + \mathbf{a}) = \mathbf{x} + \mathbf{\Lambda}\mathbf{Hc} = \mathbf{x} + \mathbf{c}, \qquad (2.43)$$

where $\mathbf{\Lambda}\mathbf{H} = \mathbf{I}$. The physical meaning of $\mathbf{c}$ is the injected offset on the system states (i.e., voltage phase angles in DC power flow model). Then, the Frobenius norm of the normalized measurement residual with false data injected $\|\overline{\pmb{\gamma}}_a\|$ is given by Liu et al. [11]

$$\begin{aligned}
\|\overline{\boldsymbol{\gamma}}_a\| &= \|\sqrt{\mathbf{W}^{-1}}(\mathbf{z_a} - \mathbf{H}\hat{\mathbf{x}}_\mathbf{a})\| \\
&= \|\sqrt{\mathbf{W}^{-1}}[\mathbf{z} + \mathbf{a} - \mathbf{H}(\hat{\mathbf{x}} + \mathbf{c})]\| \\
&= \|\sqrt{\mathbf{W}^{-1}}[\mathbf{z} - \mathbf{H}\hat{\mathbf{x}} + (\mathbf{a} - \mathbf{Hc})]\| \\
&= \|\sqrt{\mathbf{W}^{-1}}(\mathbf{z} - \mathbf{H}\hat{\mathbf{x}})\| \le \tau.
\end{aligned} \tag{2.44}$$

As we see, in this case, no anomaly can be observed by the existing bad data detection approach, which indicates a success of an FmDI attack.

Note that to construct successful FmDI attacks, the attackers must have valuable knowledge of the targeted power grid, including branch connection information, system configuration data, as well as the current system operating status. Various channels can be exploited by FmDI attackers to illegally obtain these information, including

- Cyber channels: eavesdropping, intrusion into the control center, insider theft or accidental leaks, malicious disclosure by disgruntled employees, etc.
- Physical channels: field measurements or investigation activities using specialized tools in areas with insufficient protection, and physical tampering with the hardware components of field devices.
- Cyber and physical channels: coordinated cyber intrusions and physical measurement/investigation acts.

Attackers may have various attack capabilities and, therefore, various knowledge levels of the valuable information of power grids. For example, some of them may have strong attack capabilities and remotely compromise the IED devices of interest through cyber intrusions in a short time, while some others may need a long time to gradually accumulate that information by persistent eavesdropping.

2.4 State-of-the-Art Literature

In this section, we will review the state-of-the-art literature in terms of IDSs, FDI attack detection techniques, as well as insider threat detection approaches.

2.4.1 A Taxonomy of IDSs

IDSs are one of the primary tools for the protection of computer networks, and they identify and respond to intrusion activities—entities attempting to violate security policies in place—by monitoring and analyzing system behaviors. A typical IDS is usually composed of sensors, analysis engine, and an alerting system. Sensors, deployed at different network places or hosts, are used for collecting network

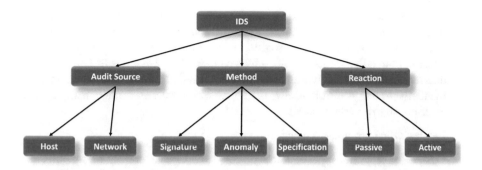

Fig. 2.2 A taxonomy of IDSs

traffic data or host behaviors. The analysis engine is responsible for data analysis with given security models, policies, or signatures. The alerting system reports to the system administrator(s) provided that an intrusion activity is identified by the analysis engine.

Generally, IDSs can be classified as host-based and network-based in terms of the audit sources; signature-based, anomaly-based, specification-based in terms of the detection methods; and passive-based and active-based in terms of the reactions [12]. A taxonomy of IDSs are provided in Fig. 2.2. Specifically,

- **Host-based**: this type of IDSs is deployed at specific hosts, e.g., sensors, substations, PCs, routers, to monitor the behaviors of such hosts malicious activities.
- **Network-based**: this type of IDSs usually connects two or more network segments to monitor the traffic over communication links in order to detect malicious intrusions.

Next, when it comes to detection methods,

- **Signature-based**: this type of IDSs detects attacks when the system or network behaviors match an attack signature/pattern stored in a database. Signature-based IDSs are effective in detecting known-signature/pattern attacks because a blacklist of known threats is stored in the database for comparison. However, this approach works ineffectively in detecting new threats or variants of known threats.
- **Anomaly-based**: this type of IDSs detects attacks when the monitored behaviors deviate from normal behaviors. Anomaly-based IDSs are effective in detecting new threats because a white list of system or network behaviors is stored in the database for comparison. However, this approach usually suffers from high false positives because any difference from the given normal behaviors is considered as being anomalous.
- **Specification-based**: this type of IDSs detects attacks when a deviation from predefined rule specifications is observed. Unlike anomaly-based approach,

in specification-based approach, specifications are defined by domain experts manually according to a set of laws, thresholds, balance requirements, etc. Specification-based approach usually has lower false positives and false negatives because manually defined specifications can avoid known cases of false positives and false negatives. However, this approach needs to define rule specifications manually to adapt different environments, which may be time-consuming and error-prone to a certain extent.

Finally, when it comes to reactions,

- **Passive-based**: this type of IDSs detects attacks that are aimed at obtaining information or monitoring the network. The main purpose of these IDSs is to notify the system operator and then provide information about attacks.
- **Active-based**: this type of IDSs detects attacks that affect the target media in the aspect of information, performance, etc. The main goal of these IDSs, at difference with passive-based IDSs, is to minimize the damage done by the attacker or locate the attacker.

After review of state-of-the-art literature, we summarize the recent studies relating to IDSs in terms of the detection method in Table 2.1. Specifically, Bao et al. in 2016 proposed a specification-based IDS to detect insider threats in smart grid CPS [2]. Thanigaivelan et al. in 2016 presented a distributed anomaly-based IDS for detecting routing attacks for Internet-of-Things (IoT) [13]. Pan et al. in 2015 developed a hybrid of signature- and specification-based IDS against multiple conventional cyberattacks as well as system disturbances [14]. Faisal et al. in 2015 proposed an anomaly-based IDS against multiple conventional attacks for advanced metering architecture in smart grid [15]. Lee et al. in 2014 developed an anomaly-based IDS against DoS attacks in 6LoWPAN (IPv6 over Low-Power Wireless Personal Area Networks) [16]. Oh et al. in 2014 proposed a signature-based IDS against multiple conventional attacks for the IoT [17].

Table 2.1 A summary of example IDSs in recent years

Reference	Security threat	Detection method
Bao et al. [2]	Insider threat	Specification-based
Thanigaivelan et al. [13]	Routing attack	Anomaly-based
Pan et al. [14]	Multiple conventional attacks and system disturbance	Hybrid signature- and specification-based
Faisal et al. [15]	Multiple conventional attacks	Anomaly-based
Lee et al. [16]	DoS	Anomaly-based
Oh et al. [17]	Multiple conventional attacks	Signature-based

2.4.2   FDI Attacks Detection

In the following two subsections, we will particularly introduce the state-of-the-art literature relating to FDI attacks detection and insider threat detection—the major focuses in this monograph. The FDI (false data injection) attack is also referred to as or similar to data deception attack, data integrity attack in the existing literature. FDI attacks are crucial security threats to smart grid CPSs, where the attackers attempt to inject false measurement data through compromised meter devices to blind and mislead the control centers. The success of an FDI attack may cause system disturbances, power overloading, power outages, and even system disruptions.

Conventional bad data detection approaches are highly dependent on the power system state estimation. Unfortunately, Liu et al. in 2011 demonstrated that new security threats can circumvent the traditional state estimation and bad data detection approach, as long as the attacker has knowledge of power system connection and configuration information and a set of compromised meter devices [11]. As a result, bad data detection approaches based on state estimation may no longer be efficient or effective anymore. This type of FDI attack is also referred to as false measurement data injection (FmDI) attacks. In terms of the data content, another type of FDI attack is false command data injection (FcDI) attacks. Since FcDI attacks are usually initiated by compromised insiders, we will further review it in the next subsection. In this subsection, we mainly introduce recent studies focusing on FmDI attack detection. Yang et al. in 2014 studied the optimal FmDI attack strategy to cause maximum damage by identifying the optimal meter set and developed the spatial-based and temporal-based schemes to detect FmDI attacks [18]. Huang et al. in 2013 reviewed the FmDI attacks as well as the defense solutions in smart grid [19]. Esmalifalak et al. in 2012 investigated the effect of stealthy FmDI attacks on network congestion in market based power system [20]. Also, Xie et al. in 2010 introduced the FmDI attacks in deregulated electricity markets and how such attacks can lead to changes in the locational marginal price and obtain illicit profits [21]. Huang et al. in 2011 proposed an adaptive CUSUM algorithm to defend FmDI attacks on smart grid network [22]. Esmalifalak et al. in 2011 used independent component analysis to detect stealth FmDI attacks where the attackers are without prior knowledge of the power grid topology [23]. In 2015, Li et al. proposed a quickest sequential detector based on the generalized likelihood ratio to detect FmDI attacks with various attack strategies [24]. Also in 2015, Chaojun et al. introduced a new detection method against FmDI attacks in smart grid by tracking the dynamics, indicated by the Kullback–Leibler distance, of measurement variations [25].

2.4.3   Insider Threats Detection

Cybersecurity is vital to the success of a smart grid CPS. The major security threats are coming from within, as opposed to outside forces. Insider threat detection is a significant part of cyber threat mitigation techniques. Insider threats are more challenging compared to outsider threats because insiders are usually empowered with legal access and privileges. Each insider has a system role associated with his/her account, such as system administrator, advanced user, normal user. Various levels of access and privileges are provided for different roles. The motivations of insider threats include

- compromised insiders
- misoperations
- emotional-driven (e.g., anger, stress, hostility)
- profit-driven

Motivated insiders either intentionally or unintentionally may cause damages to the system, delay or compromise the services, or steal or leak intellectual information. Particularly, FcDI attack is a type of specific insider threats, where, e.g., in a smart grid CPS, the attackers attempt to issue fake commands to the system actuators such as generators, breakers, substations in the purpose of causing power outages, overloading, system disturbances, or undesired changes of electricity prices. A myriad of studies focusing on insider threat detection have been observed over the years. Chen et al. in 2012 introduced an unsupervised learning based community anomaly detection system against anomalous insiders in collaborative information systems [26]. Ambre and Shekokar in 2015 proposed a probabilistic approach to detect insider threats by using log analysis and event correlation [27]. Legg et al. described an automated insider threat detection system by using user tree-structure profiling approach [28]. In 2012, Brdiczka et al. proposed a proactive insider threat detection system based on graph learning and psychological context approach [29]. In 2015, Mayhew et al. designed an enhanced behavior-based access control technique by integrating machine learning techniques against insider threat detection in big data analytics [30]. Ring et al. proposed a new toolset for anomaly-based IDS, particularly for insider threat detection [31].

References

1. Schweppe, F. C., & Rom, D. B. (1970). Power system static-state estimation, part II: Approximate model. *IEEE Transactions on Power Apparatus and Systems, 1*, 125–130.
2. Bao, H., Lu, R., Li, B., & Deng, R. (2015). BLITHE: Behavior rule based insider threat detection for smart grid. *IEEE Internet of Things Journal, 3*(2), 190–205.
3. Tian, J., Tan, R., Guan, X., & Liu, T. (2017). Hidden moving target defense in smart grids. In *Proceedings of the Second Workshop on Cyber-Physical Security and Resilience in Smart Grids, Pittsburgh* (pp. 21–26).
4. Liu, C., Wu, J., Long, C., & Kundur, D. (2018). Reactance perturbation for detecting and identifying FDI attacks in power system state estimation. *IEEE Journal on Selected Topics in Signal Processing, 12*(4), 763–776.
5. Tian, J., Tan, R., Guan, X., & Liu, T. (2018). Enhanced hidden moving target defense in smart grids. *IEEE Transactions on Smart Grid, 10*(2), 2208–2223.

6. Purchala, K., Meeus, L., Van Dommelen, D., & Belmans, R. (2005). Usefulness of DC power flow for active power flow analysis. In *Proceedings of the IEEE 2005 Power Engineering Society General Meeting, San Francisco* (pp. 454–459). Piscataway, NJ: IEEE.
7. Van Hertem, D., Verboomen, J., Purchala, K., Belmans, R., & Kling, W. L. (2006). Usefulness of DC power flow for active power flow analysis with flow controlling devices. In *Proceedings of the Eighth IEE International Conference on AC and DC Power Transmission (ACDC 2006)* (pp. 58–62). London: IET.
8. Monticelli, A. (2012). *State estimation in electric power systems: A generalized approach.* Berlin: Springer.
9. Nor, N. M., Jegatheesan, R., & Nallagownden, P. (2008). Newton–Raphson state estimation solution employing systematically constructed Jacobian matrix. In *Advanced technologies.* IntechOpen. https://doi.org/10.5772/8200.
10. Abur, A., & Exposito, A. G. (2004). *Power system state estimation: Theory and implementation.* Boca Raton, FL: CRC Press.
11. Liu, Y., Ning, P., & Reiter, M. K. (2011). False data injection attacks against state estimation in electric power grids. *ACM Transactions on Information and System Security, 14*(1), 13.
12. Zarpelão, B. B., Miani, R. S., Kawakani, C. T., & de Alvarenga, S. C. (2017). A survey of intrusion detection in Internet of Things. *Journal of Network and Computer Applications, 84,* 25–37.
13. Thanigaivelan, N. K., Nigussie, E., Kanth, R. K., Virtanen, S., & Isoaho, J. (2016). Distributed internal anomaly detection system for Internet-of-Things. In *Proceedings of the 13th IEEE Annual Consumer Communications and Networking Conference (CCNC), Las Vegas* (pp. 319–320).
14. Pan, S., Morris, T., & Adhikari, U. (2015). Developing a hybrid intrusion detection system using data mining for power systems. *IEEE Transactions on Smart Grid, 6*(6), 3104–3113 (2015)
15. Faisal, M. A., Aung, Z., Williams, J. R., & Sanchez, A. (2015). Data-stream-based intrusion detection system for advanced metering infrastructure in smart grid: A feasibility study. *IEEE Systems Journal, 9*(1), 31–44.
16. Lee, T.-H., Wen, C.-H., Chang, L.-H., Chiang, H.-S., & Hsieh, M.-C. (2014). A lightweight intrusion detection scheme based on energy consumption analysis in 6LowPAN. In *Advanced technologies, embedded and multimedia for human-centric computing* (pp. 1205–1213). Berlin: Springer.
17. Oh, D., Kim, D., & Ro, W. W. (2014). A malicious pattern detection engine for embedded security systems in the Internet of Things. *Sensors, 14*(12), 24188–24211.
18. Yang, Q., Yang, J., Yu, W., An, D., Zhang, N., & Zhao, W. (2014). On false data-injection attacks against power system state estimation: Modeling and countermeasures. *IEEE Transactions on Parallel and Distributed Systems, 25*(3), 717–729.
19. Huang, Y., Esmalifalak, M., Nguyen, H., Zheng, R., Han, Z., Li, H., & Song, L. (2013). Bad data injection in smart grid: Attack and defense mechanisms. *IEEE Communications Magazine, 51*(1), 27–33.
20. Esmalifalak, M., Han, Z., & Song, L. (2012). Effect of stealthy bad data injection on network congestion in market based power system. In *Proceedings of the 2012 IEEE Wireless Communications and Networking Conference (WCNC), Paris* (pp. 2468–2472).
21. Xie, L., Mo, Y., & Sinopoli, B. (2010). False data injection attacks in electricity markets. In *Proceedings of the First IEEE International Conference on Smart Grid Communications (SmartGridComm), Gaithersburg* (pp. 226–231).
22. Huang, Y., Li, H., Campbell, K. A., & Han, Z. (2011). Defending false data injection attack on smart grid network using adaptive CUSUM test. In *Proceedings of the 45th Annual Conference on Information Sciences and Systems (CISS), The John Hopkins University, Baltimore* (pp. 1–6).

23. Esmalifalak, M., Nguyen, H., Zheng, R., & Han, Z. (2011). Stealth false data injection using independent component analysis in smart grid. In *Proceedings of the IEEE Second International Conference on Smart Grid Communications (SmartGridComm), Brussels* (pp. 244–248).
24. Li, S., Yilmaz, Y., & Wang, X. (2015). Quickest detection of false data injection attack in wide-area smart grids. *IEEE Transactions on Smart Grid, 6*(6), 2725–2735.
25. Chaojun, G., Jirutitijaroen, P., & Motani, M. (2015). Detecting false data injection attacks in AC state estimation. *IEEE Transactions on Smart Grid, 6*(5), 2476–2483.
26. Chen, Y., Nyemba, S., & Malin, B. (2012). Detecting anomalous insiders in collaborative information systems. *IEEE Transactions on Dependable and Secure Computing, 9*(3), 332–344.
27. Ambre, A., & Shekokar, N. (2015). Insider threat detection using log analysis and event correlation. *Progress in Computer Science, 45*, 436–445.
28. Legg, P. A., Buckley, O., Goldsmith, M., & Creese, S. (2017). Automated insider threat detection system using user and role-based profile assessment. *IEEE Systems Journal, 11*(2), 503–512.
29. Brdiczka, O., Liu, J., Price, B., Shen, J., Patil, A., Chow, R., et al. (2012). Proactive insider threat detection through graph learning and psychological context. In *Proceedings of the IEEE Symposium on Security and Privacy Workshops, San Francisco*, 2012, pp. 142–149.
30. Mayhew, M. J., Atighetchi, M., Adler, A., & Greenstadt, R. (2015). Use of machine learning in big data analytics for insider threat detection. In *Proceedings of the 34th IEEE Military Communications Conference (MILCOM), Tampa* (pp. 915–922).
31. Ring, M., Wunderlich, S., Grüdl, D., Landes, D., & Hotho, A. (2017). A toolset for intrusion and insider threat detection. In *Data Analytics and Decision Support for Cybersecurity* (pp. 3–31). Berlin: Springer.

Chapter 3
SPNTA: Reliability Analysis Under Topology Attacks—A Stochastic Petri Net Approach

Building an efficient, intelligent, and multi-functional power grid with high security and reliability is a very challenging task, especially in the evolving network threat environment. The increasing complexity of the grid in the network and physical areas also exacerbates the challenges. In this chapter, we develop an analysis model based on stochastic Petri net (SPN) to evaluate and analyze the reliability of smart grid, especially for topology attacks and system countermeasures (i.e., IDSs and malfunction recovery methods). The topology attack evolved from FmDI attack is more and more threatening to the security of smart grid. In our analysis model, two types of topology attacks, namely conservative topology attacks and aggressive topology attacks, are defined and considered, then, two kinds of unreliable results, i.e., system disturbance and failure, are also considered. Taking IEEE 14 node power system as an example, the establishment and parameterization of the model are described in detail. The advantage of using this analytical model is that the system reliability can be measured from both transient analysis and steady-state analysis. Lastly, a large number of experiments are carried out to verify the feasibility as well as the effectiveness of the developed model.

3.1 Introduction

Smart grid is regarded as a revolutionary upgrade to traditional grid, whose main purpose is to enhance situational awareness of large and often dispersed physical infrastructure [1–3].

In addition, intelligent functions such as smart demand response, self-healing, self-recovery, and adaptation to renewable resources can be realized by making full use of information and communication technology [4–6]. Smart grid has attracted the interests of policy-makers and government agencies, which can be evidenced by various policies as well as initiatives released in the past few years [7–9].

© Springer Nature Switzerland AG 2020
B. Li et al., *Detection of False Data Injection Attacks in Smart Grid Cyber-Physical Systems*, Wireless Networks,
https://doi.org/10.1007/978-3-030-58672-0_3

Considering those promising benefits proposed in Sect. 1.2.2, there are still many latent risks and challenges [3, 10, 11]. System reliability affected by various grid components is one of the key problems in the smart grid [4, 12]. What is more, maintaining grid reliability needs demand-response strategy and peak shaving technology which play an important role in balancing power request and generation capacity. Other significant considerations for maintaining grid reliability include the property and life of substations, electrical equipment, and transmission lines. Renewable resources, such as water, solar, wind, hydraulic, and tidal energy, affect the reliability of the system owing to their unstable nature. Similar to other user technologies, ensuring the integrity and authenticity of measurement data reported by sensing devices is also vital to assuring grid reliability, such as system state estimation. For example, a declinational or fabricated measurement may cause the system control center to issue a wrong feedback command, thus affecting the reliability of the system.

With smart grid becoming the target of network attack and physical damage, the reliability of smart grid is affected [13, 14]. It is important to ensure the flexibility and reliability of the system design. Note that a successful compromise or attack can make a great difference, which is shown by recent events (e.g., Stuxnet [13]). Although people are interested in the research of the security of smart grid recently, the existing literature usually focuses on single event attack instead of cooperative attack. This is partly because the existing modeling and analyzing mathematical tools aiming at cooperative attacks are not effective to deal with complex cooperative attacks [15]. For example, attack tree is a commonly used tool in the state-of-the-art literature to depict the concept map of a single attack. However, most of the existing attack tree models are not good at modeling concurrent and cooperative attacks, especially when there are defensive measures. In addition, only a few researches introduce modeling tools, which can fully capture the dynamic between attack and defense, as same as the comprehensive characteristics in smart grid network physical system [16]. This is the discrepancy that will be addressed in this chapter.

Additionally, the topology attacks [17] will be introduced, a typical example of cooperative attack in smart grid environment. We then use a SPN [18, 19] to model the network topology attack and appraise the system reliability under IDSS and fault recovery technology. Topology attack is evolved from bad data injection attack, which is a research hotspot in decades. For instance, Liu et al. represented in 2011 that by destroying a set of measurement devices, an attacker can construct an attack vector and easily bypass the traditional bad data detector; therefore, a triumphant bad data injection attack is launched [20]. A key constraint to the success of such attacks is the need for large quantities of measuring equipment. This is a very strong assumption //because ordinary attackers usually have limited time and ability to build FmDI attacks in the smart grid. In order not to be trapped by the limitations related to bad data injection attacks, a new type of topology attacks has recently emerged rapidly and reduced the requirements for constructing attacks. Theoretically, if a very small group of sensing devices are damaged at the same time, the adversary can launch an effective topology attack. In addition, Petri nets are widely used in various asynchronous and concurrent process modeling tools, so they are more suitable for collaborative topology attack modeling and concurrent behavior capture of network and physical process in smart grid.

The contributions of this chapter are as follows:

- First, we propose a new analysis model to carefully evaluate and analyze the reliability under topology attacks as well as the countermeasures in smart grids (i.e., IDSs and malfunction recovery methods). Because topology attacks are generally considered "undetectable" attacks, recognizing their behavior and their potential impacts can help build more resilient and reliable smart grids.
- Second, two kinds of topology attacks in smart grids, i.e., conservative and active topology attacks, are defined and considered. We also consider the impact of different attacks and their actions on the smart grid.
- Third, we describe a scheme to identify whether an undetected damaged sensor device can initiate a successful topological attack or not. After that, the different types of effects of successful attacks, such as system interference or failure, are discussed. Furthermore, an algorithm for constructing the maximum spanning tree (MxST) [21] is also proposed for smart grids.

To the best of our knowledge, this is the first time SPN has been used in the smart grid context to study topology attacks. The choice of SPN is due to its capability of incorporating features of both the cyber domain (e.g., cyber intrusion process and corresponding state transition process) and the physical domain (e.g., physical measurement data and possible impacts and outages).

The rest of this chapter is organized as follows: Section 3.2 gives some preliminaries, then, the system model, threat model, and our design goals are described in Sect. 3.3. The details of our proposed analytical model are presented in Sect. 3.4, followed by the numerical results in Sect. 3.5. Section 3.6 concludes the chapter.

3.2 Preliminaries

In this section, we introduce some preliminaries relating to Petri net modeling as well as coordinated attacks.

3.2.1 Petri Net Modeling

Petri Net modeling technology is more and more respected by scholars, one of the reasons is due to the fast development of networks and distributed systems in recent years [19]. We describe a basic Petri net by a 4-tuple $(\mathcal{P}, \mathcal{T}, \mathcal{F}, \mathcal{M})$, where $\mathcal{P}$ denotes a finite set of places (i.e., the states), $\mathcal{T}$ denotes a finite set of transitions (that is the actions or behaviors), $\mathcal{F} \subset (\mathcal{P} \times \mathcal{T}) \cup (\mathcal{T} \times \mathcal{P})$ denotes a finite set of input arcs and output arcs, and $\mathcal{M}$ is a finite set of markings denoting the distribution of the number of tokens located in each place. Particularly, a token denotes an object holding the occurrence of a particular condition or an event. Tokens can be moved from one place to another when a particular condition changes or an event occurs. Owing to the limitation that a basic Petri net can only model a fairly simple process, a number of extended Petri nets have been recently presented in the literature to enable a wider range of applications [22].

Petri Net was first used by McDermott to model network attacks as an alternative to traditional attack trees [23]. The results show that Petri nets can describe parallel processes more effectively than traditional attack trees. Subsequently, Bause et al. introduced generalized SPN techniques to simulate network attacks [24]. A timed Petri net is nominated as SPN in which the transformation between trigger times is assumed to be exponentially distributed. Without the SPN, I could easily turn the state transition process into a reachable, which in turn turned into a continuous time Markov chain, helping the system administrator perform steady-state analysis. SPN has been increasingly adopted by the research communities and has been exploited to enable a variety of applications [25–29]. Another extension to the Petri net is the colored Petri net, where the difference is that the tokens are described by diverse colors. Unlike basic Petri nets, colored Petri nets can be leveraged to model more complicated systems or processes. For instance, Jensen suggests that colored Petri nets can be widely utilized in real-world applications, e.g., ISDN networks, naval ship systems, and also ATM networks [26].

Additionally, we use Petri net modeling techniques to describe the state transition process in power systems of communications infrastructure and physical systems [27–29]. For example, Laprie et al. used Petri nets to model interdependent faults in power infrastructure and interconnected information systems. Furthermore, Zeng et al. used SPN to analyze the reliability of the control center network in a smart grid, considering that the reliability of the network could be compromised by Byzantine failures and server failures in the control center network. However, there are no effective modeling techniques to model topology attacks in smart grids, particularly in cases of system confrontation.

3.2.2   Coordinated Attacks

In a preconceived, large-scale network physical infrastructure, a single attacker is usually unable to disrupt the general infrastructure. The CERT report [30] shows that well-resourced attackers, such as state-sponsored actors and organized cybercrime groups, are in most cases more likely to attempt collective, coordinated attacks.

In smart grid CPS environment, cooperative attack includes false data injection attack [3, 10, 20], topology attack [17, 31], DoS attack, and DDoS attack [32, 33]. Because Liu has demonstrated that an attack launched by the coordinated attackers can effectively bypass the conventional bad data detection mechanisms in power systems [20], researchers have begun to study such attacks, such as [10] proposed an attack strategy investigation, potential impact on power systems, and potential countermeasures. By destroying a set of ideal data meters, an attack vector encapsuled with injected false measurement data can be constructed. If designed properly, injected attack vectors can easily bypass bad data detectors in the data center without triggering alarms. Hug et al. have presented a new state estimation vulnerability analysis technique developed to identify hidden false data injection attacks [34]. A collaborative detection scheme based on distributed host is proposed for the attack of false data injection in smart grid network physical system [3].

This chapter focuses on topological attacks which are not accessible widely. Topology attacks are regarded to be evolutionary spurious data injection attacks where measurement data and circuit breaker status data used to determine the current system topology need to be manipulated. Similar to the false data injection attacker, the topology attacker attempts to blind the broken data detector by constructing matching instrument data and circuit breaker status data. Some existing research has discussed this kind of attack. For instance, Weimer et al. has come up with a distributed way to detect and isolate grid topology attacks [31]. Kim and Tong proposed a graph-based approach to counter topological attacks by placing phasor measurement units optimally on the grid [17]. In addition to the few studies on topology attacks mentioned above, there is another particularly relevant area that has not been fully studied, namely topology attack modeling and system reliability analysis when subjected to such attacks. Therefore, in this chapter, we provide a smart grid analysis model based on SPN to model the attack behavior of topology attacks, as well as the system reliability under such attacks.

3.3 Models and Design Goals

In this section, we elaborate on our system model, threat model, and the design goals.

3.3.1 System Model

In this chapter, the state estimation workflow in power system is proposed as our system model (see Fig. 3.1). We can find the basic concepts of state estimation in Sect. 2.1. In this section, we further measure the data and enter it into a state estimator, a topology processor, and a bad data detector.

In the state estimation, two types of data are collected by the system control center in the sensing devices of the whole grid. One data type is nominated as the line flow and nodal injection analog measurement data $\mathbf{z} = \{\mathbf{P}_i, \mathbf{Q}_i, \mathbf{P}_{ij}, \mathbf{Q}_{ij}, \mathbf{V}, \boldsymbol{\theta}\}$ provided by line meters, where $\mathbf{P}_i, \mathbf{Q}_i, \mathbf{P}_{ij}, \mathbf{Q}_{ij}, \mathbf{V}, \boldsymbol{\theta}$ denote, respectively, the real power injections, reactive power injections, real power flows, reactive power flows, bus voltage magnitudes, and bus voltage angles. In addition, the circuit breaker on/off status data $\mathbf{s} = s_i^N$ is another type of data, where $s_i \in \{0, 1\}$ and N denotes the total number of branches in a smart grid. The circuit breaker monitors can

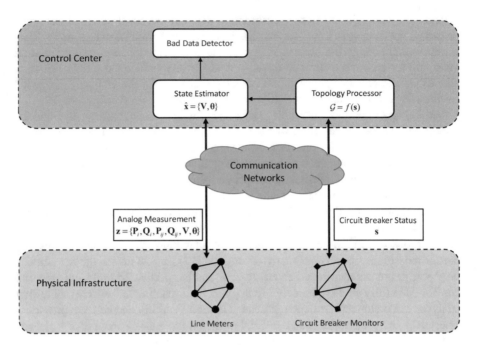

Fig. 3.1 System model: power system state estimation

provide the status data $\mathbf{s}$ [35]. After that, the current grid topology $\mathcal{G}$ is determined through analyzing $\mathbf{s}$ with the topology processor, that is, $\mathcal{G} = f(\mathbf{s})$. To further process data, both grid topology $\mathcal{G}$ and measurement data $\mathbf{z}$ are delivered to the state estimator, which then creates the estimated real system status data $\mathbf{x} = \{\hat{\mathbf{V}}, \hat{\boldsymbol{\theta}}\}$, making advantage of an AC or DC power flow model. In the final step, the bad data detector decides whether or not any bad data has been reported by the sensing devices through residual check.

As mentioned in Sect. 2.1, the correlation between the real system status data $\mathbf{x}$ and measurement data $\mathbf{z}$ based on the DC power flow model is given by [10]

$$\mathbf{z} = \mathbf{H}_{\mathcal{G}}\mathbf{x} + \eta, \tag{3.1}$$

where, particularly, $\mathbf{H}_{\mathcal{G}} \in \mathbb{R}^{m \times n}$ denotes the measurement Jacobian matrix which combines with the current system topology $\mathcal{G}$. The estimated $\hat{\mathbf{x}}$ is given by

$$\hat{\mathbf{x}} = \arg\min_{\mathbf{x}}(\mathbf{z} - \mathbf{H}_{\mathcal{G}}\mathbf{x})^{\mathsf{T}}\mathbf{W}^{-1}(\mathbf{z} - \mathbf{H}_{\mathcal{G}}\mathbf{x}) = (\mathbf{H}_{\mathcal{G}}^{\mathsf{T}}\mathbf{W}^{-1}\mathbf{H}_{\mathcal{G}})^{-1}\mathbf{H}_{\mathcal{G}}^{\mathsf{T}}\mathbf{W}^{-1}\mathbf{z} \triangleq \mathbf{\Lambda}\mathbf{z}, \tag{3.2}$$

where

$$\mathbf{\Lambda} \triangleq (\mathbf{H}_{\mathcal{G}}^{\mathsf{T}}\mathbf{W}^{-1}\mathbf{H}_{\mathcal{G}})^{-1}\mathbf{H}_{\mathcal{G}}^{\mathsf{T}}\mathbf{W}^{-1}. \tag{3.3}$$

Then, we can get the estimated measurement data $\hat{\mathbf{z}}$ through

$$\hat{\mathbf{z}} = \mathbf{H}_{\mathcal{G}}\hat{\mathbf{x}} = \mathbf{H}_{\mathcal{G}}\Lambda\mathbf{z}, \tag{3.4}$$

and the measurement residual $\boldsymbol{\gamma}$ is calculated by

$$\boldsymbol{\gamma} = \mathbf{z} - \hat{\mathbf{z}} = \mathbf{z} - \mathbf{H}_{\mathcal{G}}\Lambda\mathbf{z} = (\mathbf{I} - \mathbf{H}_{\mathcal{G}}\Lambda)\mathbf{z}. \tag{3.5}$$

3.3.2 Adversary Model

In this chapter, we think about the situation where the opponents are topology attackers. The purpose of this is to reduce the reliability and survivability of the system. As we proposed earlier, topology attacks can be viewed as evolutionary FmDI attacks. With respect to spurious data injection attacks, opponents typically tamper only with metric $\mathbf{z}$. In practice, because of a mismatch with the system topology $\mathcal{G}$, bad data detector can usually detect these fake data. Anyway, the attackers of topology are aiming to fake both circuit breaker status data $\mathbf{s}$ corresponding to the system topology $\mathcal{G}$ (i.e., $\mathcal{G} = f(\mathbf{s})$ and measurement data $\mathbf{z}$ [17]. Generally speaking, they are going to construct a pair of matched measurements with a grid topology, where such an attack strategy can be available in spoofing bad data detectors.

The adversary model is shown in Fig. 3.2, we presume that the sensing devices, including the line meters and circuit breaker monitors in smart grids, can be compromised by malicious adversaries, both internal and external. The attackers are able to report false measurement data, by the control of the sensing devices. In Fig. 3.2, we construct attack vectors $\mathbf{a}$ and $\mathbf{a}_s$, denoted by red circles and squares, respectively. Then, vector $\mathbf{z}_{\mathbf{a}}$ and $\mathbf{s}_{\mathbf{a}}$ can be given as below:

$$\mathbf{z}_{\mathbf{a}} = \mathbf{z} + \mathbf{a}, \tag{3.6}$$

and

$$\mathbf{s}_{\mathbf{a}} = \mathbf{s} + \mathbf{a}_s. \tag{3.7}$$

Accordingly, the processed grid topology is composed of

$$\mathcal{G}_{\mathbf{a}} = f(\mathbf{s}_{\mathbf{a}}) = f(\mathbf{s} + \mathbf{a}_s), \tag{3.8}$$

where $f(\cdot)$ denotes a function representing the topology processor.

Similarly, the relation between $\mathbf{z}_{\mathbf{a}}$ and $\mathbf{x}$ is expressed as follows, taking the DC power flow model into account:

$$\mathbf{z}_{\mathbf{a}} = \mathbf{H}_{\mathcal{G}_{\mathbf{a}}}\mathbf{x} + \eta. \tag{3.9}$$

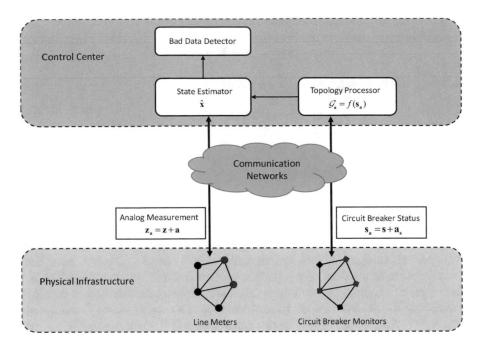

Fig. 3.2 Adversary model

In addition, the estimated $\hat{\mathbf{x}}_{\mathbf{a}}$ can be given by

$$
\begin{aligned}
\hat{\mathbf{x}}_{\mathbf{a}} &= \arg\min_{\mathbf{x}}(\mathbf{z}_{\mathbf{a}} - \mathbf{H}_{\mathcal{G}_{\mathbf{a}}}\mathbf{x})^{\mathsf{T}}\mathbf{W}^{-1}(\mathbf{z}_{\mathbf{a}} - \mathbf{H}_{\mathcal{G}_{\mathbf{a}}}\mathbf{x}) \\
&= (\mathbf{H}_{\mathcal{G}_{\mathbf{a}}}^{\mathsf{T}}\mathbf{W}^{-1}\mathbf{H}_{\mathcal{G}_{\mathbf{a}}})^{-1}\mathbf{H}_{\mathcal{G}_{\mathbf{a}}}^{\mathsf{T}}\mathbf{W}^{-1}\mathbf{z}_{\mathbf{a}} \triangleq \mathbf{\Lambda}_a\mathbf{z}_{\mathbf{a}},
\end{aligned}
\tag{3.10}
$$

where

$$
\mathbf{\Lambda}_a = (\mathbf{H}_{\mathcal{G}_{\mathbf{a}}}^{\mathsf{T}}\mathbf{W}^{-1}\mathbf{H}_{\mathcal{G}_{\mathbf{a}}})^{-1}\mathbf{H}_{\mathcal{G}_{\mathbf{a}}}^{\mathsf{T}}\mathbf{W}^{-1}.
\tag{3.11}
$$

Then, the estimated measurement data $\hat{\mathbf{z}}_{\mathbf{a}}$ is calculated by

$$
\hat{\mathbf{z}}_{\mathbf{a}} = \mathbf{H}_{\mathcal{G}_a}\hat{\mathbf{x}} = \mathbf{H}_{\mathcal{G}_{\mathbf{a}}}\mathbf{\Lambda}_a\mathbf{z}_{\mathbf{a}}.
\tag{3.12}
$$

The normalized residual $\overline{\boldsymbol{\gamma}}_a$ is then given by

$$
\overline{\boldsymbol{\gamma}}_a = \sqrt{\mathbf{W}^{-1}}(\mathbf{z}_{\mathbf{a}} - \hat{\mathbf{z}}_{\mathbf{a}}) = \sqrt{\mathbf{W}^{-1}}(\mathbf{I} - \mathbf{H}_{\mathcal{G}_{\mathbf{a}}}\mathbf{\Lambda}_a)\mathbf{z}_{\mathbf{a}}.
\tag{3.13}
$$

The final key step is to identify the falsified data. The Frobenius norm $\|\overline{\boldsymbol{\gamma}}_a\| = \|\sqrt{\mathbf{W}^{-1}}(\mathbf{I} - \mathbf{H}_{\mathcal{G}_{\mathbf{a}}}\mathbf{\Lambda}_a)\mathbf{z}_{\mathbf{a}}\|$ can be regarded as a function of $\mathbf{a}$ and $\mathbf{s}_{\mathbf{a}}$ (recall that

$\mathcal{G} = f(\mathbf{s})$). Using this method, the constructed vectors $\mathbf{a}$ and $\mathbf{s_a}$ can result in

$$\|\overline{\boldsymbol{\gamma}_a}\| = \|\sqrt{\mathbf{W}^{-1}}(\mathbf{I} - \mathbf{H}_{\mathcal{G}_a}\boldsymbol{\Lambda}_a)\mathbf{z_a}\| < \tau, \qquad (3.14)$$

and the opponents can launch available topology attacks.

According to different attack strategies, two categories of topology attacks are defined in this chapter, which are shown as below:

- *Conservative topology attacks:* the aim of this type of attacks is to manipulate one or more transmission lines or buses by destroying very few sensing devices. As a result, the manipulation of these limited resources has little influence on the current power system.
- *Aggressive topology attacks:* these attacks manipulate large areas of the grid as much as possible (for example, by destroying as many sensors as possible). These attacks, if successful, can cause systems to crash or fail, which often results in devastating damage to power systems.

It should be noted that we do not consider data integrity attacks (e.g., man-in-the-middle attacks) against communication links in the adversary/threat model for each part of the work covered in this monograph. We propose solutions to detect FDI attacks based on data integrity while ensuring the security of the data transmission networks. If necessary, the BLS short signature [36] can be employed to ensure data integrity during transmission.

3.3.3 Design Goals

The main aim of this paper is to design an analytic model for studying the topology attack in smart grid and to analyze the system reliability under these attacks. Furthermore, our design goals are as follows:

1. The attack strategies of different types of topology attacks and their potential impacts on power systems are going to be analyzed in depth.
2. A state transition model based on SPN is going to be established to describe the behavior of the system under topology attack.
3. Reliability evaluation of smart grid system based on reliability measurement is going to be defined.

3.4 Proposed Analytical Model

In this section, an analytical model based on SPN (see Fig. 3.3) is proposed to describe the existing topological attacks of the system and the countermeasures (e.g., malfunction recovery techniques and IDSs) to the system security.

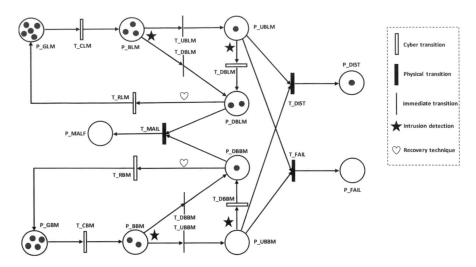

Fig. 3.3 Analytical SPN model

3.4.1   Construction of the Proposed SPN Model

The SPN model's construction is introduced in this subsection. The physical meaning of the location and transformation is annotated separately in Tables 3.1 and 3.2. The cyber transitions are represented by blank bars, while the physical transitions are denoted by filled bars. Note that, in particular, immediate transitions also appear in our model, which belong to cyber transitions and they are presented by slim vertical bars. In this model of SPN, we take two kinds of sensing devices into account, i.e., line meters and circuit breaker monitors. The sensing devices that hold specific conditions are represented using small filled red circles (tokens). With respect to countermeasures, filled black stars ★ are used to represent the presence of IDSs. The IDSs are deployed to periodically detect the sensing device malfunctions. Furthermore, to recover the malfunction devices recognized by IDSs, we design the malfunction recovery techniques, which are represented by ♡.

In the proposed SPN model, we have an eleven-element set for places, that is, $\mathcal{P} = \{$P_GLM, P_BLM, P_GBM, P_BBM, P_DBLM, P_UBLM, P_DBBM, P_UBBM, P_DIST, P_FAIL, P_MALF$\}$. To be specific, the counts for good line meters, good breaker monitors, bad breaker monitors, and bad line meters are controlled by places P_GLM, P_BLM, P_GBM, P_BBM, respectively. Likewise, the counts for detected bad line meters, undetected bad line meters, detected bad breaker monitors, and undetected bad breaker monitors are controlled by places P_DBLM, P_UBLM, P_DBBM, P_UBBM, respectively. If a token is held, place P_DIST represents a system disturbance event caused by places P_UBLM and P_UBBM when enabling transition T_DIST. Similarly, if place P_FAIL holds a

Table 3.1 Places in the SPN model

Place	Meaning
P_GLM	The place for good line meter(s)
P_BLM	The place for bad line meter(s)
P_GBM	The place for good breaker monitor(s)
P_BBM	The place for bad breaker monitor(s)
P_DBLM	The place for detected bad line meter(s)
P_UBLM	The place for undetected bad line meter(s)
P_DBBM	The place for detected bad breaker monitor(s)
P_UBBM	The place for undetected bad breaker monitor(s)
P_DIST	The place for system disturbance: 0 for before and 1 for after
P_FAIL	The place for system failure: 0 for before and 1 for after
P_MALF	The place for system malfunction: 0 for before and 1 for after

Table 3.2 Transitions in the SPN model

Transition	Meaning
T_CLM	Transition that a line meter is compromised by the attacker
T_CBM	Transition that a breaker monitor is compromised by the attacker
T_DBLM	Transition that a bad line meter is detected by the IDS
T_UBLM	Transition that a bad line meter is failed to be detected by the IDS
T_DBBM	Transition that a bad breaker monitor is detected by the IDS
T_UBBM	Transition that a bad breaker monitor is failed to be detected by the IDS
T_RLM	Transition that a line meter is recovered by the system
T_RBM	Transition that a breaker monitor is recovered by the system operator
T_DIST	Transition that the power grid experiences a system disturbance
T_FAIL	Transition that the power grid experiences a system failure
T_MALF	Transition that the power grid experiences a system malfunction

token, it represents a system failure event caused by places P_UBLM and P_UBBM when enabling transition T_FAIL. In particular, if the P_MALF position holds a token, the entire power system will be shifted from P_DBLM to P_DBBM when a failure of the line table and circuit breaker monitor is detected and the security system cannot recover in time.

The following events are used to show how we constructed the SPN model, including how the system responds to various event triggers.

- Model initialization is the first event. A token at a location is used to represent a sensor device that meets the conditions specified for that location. Particularly, it denotes the presence of this event if tokens are held by places P_DIST, P_FAIL, and P_MALF; otherwise, empty places indicate no occurrence of such events. A tag is a sequence of token states for all locations, represented by $M = \{m_{glm}, m_{blm}, m_{gbm}, m_{bbm}, m_{dblm}, m_{ublm}, m_{dbbm}, m_{ubbm}, m_{dist}, m_{fail}, m_{malf}\}$, where, particularly as abovementioned, m_{dist}, m_{fail}, and m_{malf} can only take

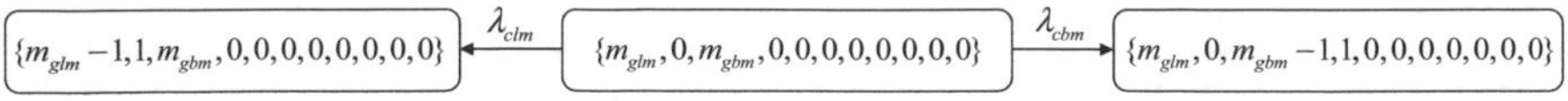

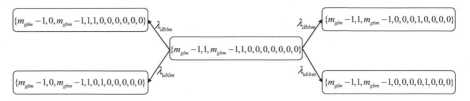

Fig. 3.4 The system state transition fired by the second event

Fig. 3.5 The system state transition fired by the third event

values of either 0 or 1. Primarily, all the devices are uncompromised/good, so we can initialize the marking to $\mathcal{M}_0 = \{m_{glm}, 0, m_{gbm}, 0, 0, 0, 0, 0, 0, 0, 0\}$.

- An attacker that compromises a line meter or a breaker monitor is the second event. The compromising rates λ_{clm} and λ_{cbm} are used to indicate the average number of tokens per token in good places that can be moved to bad places per unit of time. We can see such transitions from Fig. 3.4. Triggering the transformation moves a token from the input position to the output position. For instance, firing the transition T_CLM in state $\{m_{glm}, 0, m_{gbm}, 0, 0, 0, 0, 0, 0, 0, 0\}$ moves one token from place P_GLM to P_BLM, transferring the state to $\{m_{glm} - 1, 1, m_{gbm}, 0, 0, 0, 0, 0, 0, 0, 0\}$.

- The third event is focusing on successfully identifying or failing to identify a compromised bad sensing device from place P_BLM or P_BBM using the IDS. For a newly compromised device within a detection interval, the IDS may fire two kinds of transitions. For example, it can be seen from Fig. 3.5, if the IDS successfully detects a bad line meter in state $\{m_{glm}-1, 1, m_{gbm}-1, 1, 0, 0, 0, 0, 0, 0, 0\}$, T_DBLM will be fired transferring the system to state $\{m_{glm} - 1, 0, m_{gbm} - 1, 1, 1, 0, 0, 0, 0, 0, 0\}$ with a rate λ_{dblm}; otherwise, T_UBLM will be fired transferring the system to state $\{m_{glm}-1, 0, m_{gbm}-1, 1, 0, 1, 0, 0, 0, 0, 0\}$ with a rate λ_{ublm}. Similar transitions for bad breaker monitor transitions are also shown in Fig. 3.5.

- The fourth event is related with identifying a compromised bad device in place P_UBLM or place P_UBBM by using the IDSs. Devices fell in the place P_UBLM or P_UBBM are regarded as compromised bad devices which have not yet been detected. The IDS checks all the devices periodically, perhaps by trust reputation [1]; hence, in this case, the damaged device may be detected during any detection interval or remain undetected for a considerable period of time. As is shown in Fig. 3.6, if a bad line meter in state $\{m_{glm} - 2, 0, m_{gbm} - 1, 0, 1, 1, 0, 1, 0, 0, 0\}$ is successfully detected by the IDS, T_DBLM will be fired transferring the system to state $\{m_{glm} - 2, 0, m_{gbm} - 1, 0, 2, 0, 0, 1, 0, 0, 0\}$ with a rate λ_{dblm}. Similarly, if transition T_DBBM is fired, a bad breaker

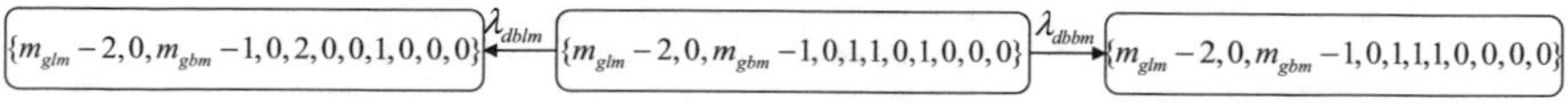

$$\{m_{glm} - 2, 0, m_{gbm} - 1, 0, 2, 0, 0, 1, 0, 0, 0\} \xleftarrow{\lambda_{dblm}} \{m_{glm} - 2, 0, m_{gbm} - 1, 0, 1, 1, 0, 1, 0, 0, 0\} \xrightarrow{\lambda_{dbbm}} \{m_{glm} - 2, 0, m_{gbm} - 1, 0, 1, 1, 1, 0, 0, 0, 0\}$$

Fig. 3.6 The system state transition fired by the fourth event

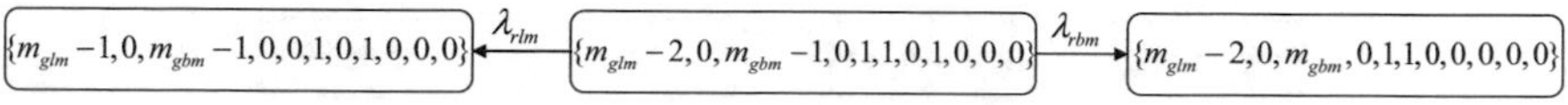

$$\{m_{glm} - 1, 0, m_{gbm} - 1, 0, 0, 1, 0, 1, 0, 0, 0\} \xleftarrow{\lambda_{rlm}} \{m_{glm} - 2, 0, m_{gbm} - 1, 0, 1, 1, 0, 1, 0, 0, 0\} \xrightarrow{\lambda_{rbm}} \{m_{glm} - 2, 0, m_{gbm}, 0, 1, 1, 0, 0, 0, 0, 0\}$$

Fig. 3.7 The system state transition fired by the fifth event

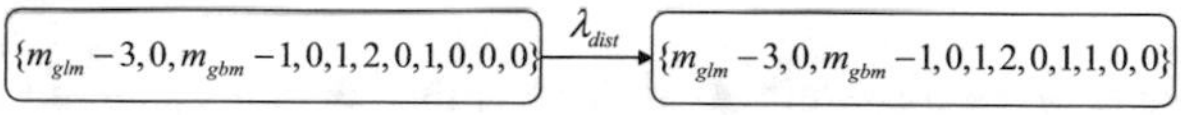

$$\{m_{glm} - 3, 0, m_{gbm} - 1, 0, 1, 2, 0, 1, 0, 0, 0\} \xrightarrow{\lambda_{dist}} \{m_{glm} - 3, 0, m_{gbm} - 1, 0, 1, 2, 0, 1, 1, 0, 0\}$$

Fig. 3.8 The system state transition fired by the sixth event

monitor will be detected transferring the system to state $\{m_{glm} - 2, 0, m_{gbm} - 1, 0, 1, 1, 1, 0, 0, 0, 0\}$ with a rate λ_{dbbm}.

- The fifth event involved the use of malfunction recovering technologies to recover a detected bad device. When a broken device is successfully detected by an intrusion detection system, the system administrator performs a malfunction recovery technique to record and reset the compromised bad device. It can be seen from Fig. 3.7, if transition T_RLM is fired with a rate λ_{rlm} in state $\{m_{glm} - 2, 0, m_{gbm} - 1, 0, 1, 1, 0, 1, 0, 0, 0\}$, an identified bad line meter is recovered to be a good line meter, transferring the system state to $\{m_{glm} - 1, 0, m_{gbm} - 1, 0, 0, 1, 0, 1, 0, 0, 0\}$. Likewise, if transition T_RBM is fired with a rate λ_{rbm} in state $\{m_{glm} - 2, 0, m_{gbm} - 1, 0, 1, 1, 0, 1, 0, 0, 0\}$, restore a detected bad breaker monitor to a good breaker monitor, transferring the system state to $\{m_{glm} - 2, 0, m_{gbm}, 0, 1, 1, 0, 0, 0, 0, 0\}$.
- The sixth event takes a successful topology attack into account, which is usually conservative and causes a system disturbance. A conservative topology attack can be constructed by a few undetected bad line meters and breaker monitors, to fire transition T_DIST. An example is shown in Fig. 3.8, where a system disturbance occurs with state $\{m_{glm} - 3, 0, m_{gbm} - 1, 0, 1, 2, 0, 1, 1, 0, 0\}$ resulting from state $\{m_{glm} - 3, 0, m_{gbm} - 1, 0, 1, 2, 0, 1, 0, 0, 0\}$ when T_DIST is enabled. The enabling function is a complex process based on the power grid topology spanning tree. The next subsection will introduce the unreliability enabling scheme that details the enabling function.
- The seventh event takes a successful topology attack into account, which is usually aggressive and causes a system failure. An aggressive topology attack can be collectively constructed by a majority of undetected bad line meters and breaker monitors, to fire transition T_FAIL. An instance is in Fig. 3.9, where a system failure happens with state $\{m_{glm} - 5, 0, m_{gbm} - 2, 0, 1, 4, 0, 2, 0, 1, 0\}$ resulting from state $\{m_{glm} - 5, 0, m_{gbm} - 2, 0, 1, 4, 0, 2, 0, 0, 0\}$ when T_FAIL is enabled. Similarly, the enabling function is a complex process based on the grid topology spanning tree, which is explained in detail in the unreliability enabling scenario described in the next section.

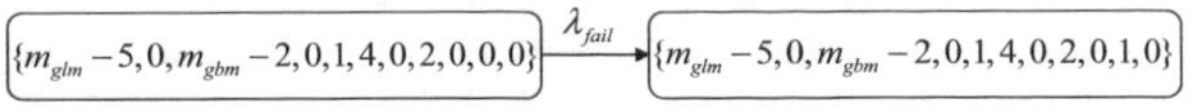

$$\{m_{glm} - 5,0,m_{gbm} - 2,0,1,4,0,2,0,0,0\} \xrightarrow{\lambda_{fail}} \{m_{glm} - 5,0,m_{gbm} - 2,0,1,4,0,2,0,1,0\}$$

Fig. 3.9 The system state transition fired by the seventh event

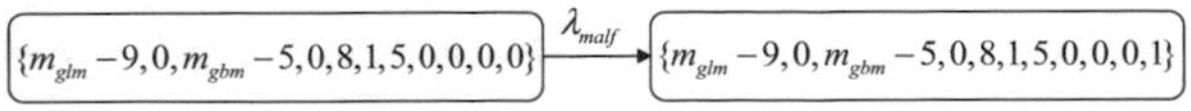

$$\{m_{glm} - 9,0,m_{gbm} - 5,0,8,1,5,0,0,0,0\} \xrightarrow{\lambda_{malf}} \{m_{glm} - 9,0,m_{gbm} - 5,0,8,1,5,0,0,0,1\}$$

Fig. 3.10 The system state transition fired by the eighth event

- The last event takes a system malfunction into account, owing to a scarce of good sensing devices. Provided that the system's recovery rate is relatively low (i.e., λ_{rlm} and λ_{rbm} are considerably small values), the detected bad devices cannot be recovered immediately, which leaves a large number of detected bad devices to remain in places P_DBLM and P_DBBM. In this case, the wide area monitoring does not have sufficient good sensing devices to work normally to support the function. After that, the power system malfunctions, because the system states are no longer fully observable to the system control center [37]. As we see, transition T_MALF can be corporately fired by a large amount of bad line meters which are unrecovered and breaker monitors to cause a system malfunction. As Fig. 3.10 shows, a system malfunction with state $\{m_{glm} - 9,0,m_{gbm} - 5,0,8,1,5,0,0,0,1\}$ may occur from state $\{m_{glm} - 9,0,m_{gbm} - 5,0,8,1,5,0,0,0,0\}$ with a rate λ_{malf} when T_MALF is enabled. The next subsection will introduce the unreliability enabling scheme where the enabling function is straightforward and integrated.

3.4.2 Maximum Spanning Tree Based Unreliability Enabling Scheme

We now show the proposed scheme that is composed of two algorithms, determining under what conditions the power system will fall into which unreliability status (i.e., disturbance, failure, or malfunction).

3.4.2.1 MxST Construction Algorithm

As we planned, the spanning tree in graph theory is used to determine the most critical measurements. A graph G may have multiple spanning trees according to the compression-deletion theorem [38]. The MxST [21] is used to obtain the best results. MxST is a spanning tree of a weighted graph G, where the weight sum of all edges is the maximum over all G's spanning trees. According to its observability and reliability, we assign weights to branches to denote the different levels of importance of the power grid. This is the process we use the grid topology of the MxST to identify the most critical branches of the grid.

Table 3.3 Weights assigned to each bus

Bus type	Description	Weight assigned
Type 1	Bus with line(s) only but no generator or load	1 unit
Type 2	Bus with line(s) and load(s) but no generator	2 units
Type 3	Bus with line(s) and generator but no load	3 units
Type 4	Bus with line(s), generator, and load(s)	4 units

Algorithm 3.1 MxST construction

Input: Initial graph $G = \{\mathcal{V}_G, \mathcal{E}_G\}$ of a power grid topology; set of weights $\mathcal{W}$ assigned for all the branches
Output: A MxST $S = \{\mathcal{V}_S, \mathcal{E}_S\}$
 1: *Initialization:* $\mathcal{V}_S = \emptyset, \mathcal{E}_S = \emptyset$
 2: Step 1: Weight Assignment for each branch.
 3:     (1.1). Assign a total weight to each bus according to Table 3.3.
 4:     (1.2). Equally divide the weight assigned for each bus into k parts, where k is the number of branches connected to this bus.
 5:     (1.3). Assign the divided weights to each connected branch.
 6:     (1.4). Add the weights for each branch assigned from the two end buses.
 7: Step 2: Arrange all branches in a decreasing order of their weights using, for example, the quicksort algorithm [40].
 8: Step 3: Add to $\mathcal{E}_S$ with ϵ_{ij} that has the maximum weight $\omega_{ij} \in \mathcal{W}$; add to $\mathcal{V}_S$ with v_i and v_j that is connected by ϵ_{ij}.
 9: Step 4: Remove v_i and v_j from $\mathcal{V}_G$, and ϵ_{ij} from $\mathcal{E}_G$.
10: Step 5: Loop over all the remaining edges $\epsilon_{kt} \in \mathcal{E}_G$ connecting to vertices $v_k \in \mathcal{V}_S$. Add the edge ϵ_{kt} has currently the maximum weight $\omega_{kt} \in \mathcal{W}$ to $\mathcal{E}_S$; add $v_t \in \mathcal{V}_G$ but $v_t \notin \mathcal{V}_S$ to $\mathcal{V}_S$.
11: Step 6: Remove the edge ϵ_{kt} from $\mathcal{E}_G$ and v_t from $\mathcal{V}_G$ concerned in the last step.
12: Step 7: Repeat Steps 5 and 6 until $\mathcal{V}_G = \emptyset$.
13: Step 8: Add all generators to $\mathcal{V}_S$, and generator and load edges to $\mathcal{E}_S$.
14:
        Return: $S = \{\mathcal{V}_S, \mathcal{E}_S\}$

For each line, we try to determine the weights in different ways. In this chapter, a simple method is used to achieve this goal, which can be described in Table 3.3. Buses have four types, namely: first, buses with line(s) only but no generator or load; second, buses with line(s) and load(s) but no generator; third, buses with line(s) and generator but no load; and last, buses with line(s), generator, and load(s) [39]. From the system administrator's perspective, the system administrator needs to pay more attention to the bus with generator and/or load due to cost and reliability concerns; buses with generators are cheaper than buses with loads. Hence, we simply assign branches connected to a bus with a total weight of 1, 2, 3, and 4 units, respectively, for the four types of buses. The total weight is then evenly distributed among all adjacent branches. For example, if a type 2 bus has 4 branches, then this bus is assigned a total weight of $\omega = 2$ units, and each of its 4 branches is allocated a weight of $\omega_i = 2/4 = 0.5$ unit, where $i \in \{1, 2, 3, 4\}$.

Using this method, an algorithm is developed to show the construction of an MxST in a power grid (see Algorithm 3.1), with reference to the existing algorithm for constructing a minimum spanning tree [41, 42]. In Algorithm 3.1, the weight of all branches is calculated by dividing the total weight of each bus equally and adding the two component weights of each branch. Then, the weighted graph is constructed progressively by MxST. Starting from the highest-weighted branches, the interested branches and buses are added to the MxST in descending weight order until all bus nodes are added to the MxST, but avoid adding them repeatedly. Finally, because of the generators and loads' importance to the grid, we put all the generators, as well as generator and load edges of the MxST. Note that the time complexity of Algorithm 3.1, which is mainly determined by the time complexity of the sorting algorithm used in Step 2, is $O(|\mathcal{E}_G| \log |\mathcal{E}_G|)$ if using the quicksort algorithm [40].

In order to prove the correctness of this algorithm, we give the following theorem with its proof.

Theorem 3.1 *After running Algorithm 3.1 on a connected weighted graph $\mathcal{G}$, its output S is an MxST.*

Proof First, S is a spanning tree. This is because:

- S is a forest. No cycles are ever created.
- S is spanning. Suppose that there is a vertex v_k that is not incident to the edges of S. Then the incident edges of v_k must have been considered in the algorithm at some step. The first edge (in edge order) would have been included because it could not have created a cycle, which contradicts the definition of S.
- S is connected. Suppose that S is not connected. Then S has two or more connected components. Since $\mathcal{G}$ is connected, then these components must be connected by some edges in $\mathcal{G}$, not in S. The first of these edges (in edge order) would have been included in S because it could not have created a cycle, which contradicts the definition of S.

Second, S is a spanning tree of maximum weight. We will prove this using induction. Let S^* be an MxST. If $S = S^*$, then S is an MxST. If $S \neq S^*$, then there exists an edge $e \in S^*$ of maximum weight that is not in S. Further, $S \cup e$ contains a cycle C such that:

1. Every edge in C has weight larger than $wt(e)$, where $wt(\cdot)$ presents the weight of an edge. (This follows from how the algorithm constructed S.)
2. There is some edge f in C that is not in S^*. (Because S^* does not contain the cycle C.)

Consider the tree $S_2 = S \setminus \{e\} \cup \{f\}$:

1. S_2 is a spanning tree.
2. S_2 has more edges in common with S^* than S did.
3. And $wt(S_2) \leq wt(S)$. (We exchanged an edge for one that is no more expensive.) We can redo the same process with S_2 to find a spanning tree S_3 with more edges in common with S^*. By induction, we can continue this process until we reach S^*, from which we see $wt(S) \geq wt(S_2) \geq wt(S_3) \geq \cdots \geq wt(S^*)$.

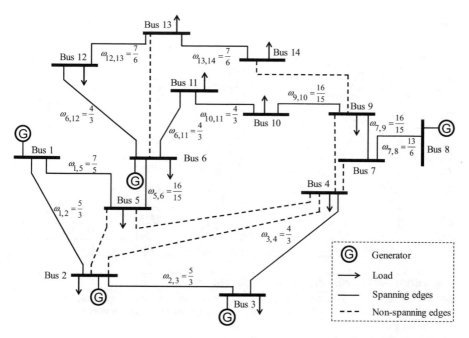

Fig. 3.11 The MxST of the IEEE 14-bus system

Table 3.4 The bus type and total weight assigned in IEEE 14-bus system

Bus	Bus type	Weight	Bus	Bus type	Weight
#1	Type 3	3 units	#8	Type 3	3 units
#2	Type 4	4 units	#9	Type 2	2 units
#3	Type 4	4 units	#10	Type 2	2 units
#4	Type 2	2 units	#11	Type 2	2 units
#5	Type 2	2 units	#12	Type 2	2 units
#6	Type 4	4 units	#13	Type 2	2 units
#7	Type 1	1 unit	#14	Type 2	2 units

Since S^* is an MxST, then these inequalities must be equalities and we conclude that S is an MxST.   □

Taking IEEE 14-bus power system (it can be seen from Fig. 3.11) as an instance to describe an MxST's construction in a power system. Based on the scheme proposed, we summarize the bus types of this power system and the total weights assigned to each bus in Table 3.4. Then, all branches' weights in the power system are calculated and listed in Table 3.5. It can be seen from Fig. 3.11, we also construct the MxST of the IEEE 14-bus system according to Algorithm 3.1, wherein all the MxST branches are represented by solid red lines and all the MxST branches' weights are annotated.

Table 3.5 Weights assigned for each branch in IEEE 14-bus system (g denotes generator and l denotes load)

Branch	Weight	Branch	Weight	Branch	Weight
$\epsilon_{1,2}$	3/3+2/3=5/3	$\epsilon_{4,7}$	2/6+2/3=1	$\epsilon_{8,g}$	3/2
$\epsilon_{1,5}$	3/3+2/5=7/5	$\epsilon_{4,9}$	2/6+2/5=11/15	$\epsilon_{9,10}$	2/5+2/3=16/15
$\epsilon_{1,g}$	3/3=1	$\epsilon_{4,l}$	2/6=1/3	$\epsilon_{9,14}$	2/5+2/3=16/15
$\epsilon_{2,3}$	4/6+4/4=5/3	$\epsilon_{5,6}$	2/5+4/6=16/15	$\epsilon_{9,l}$	2/5
$\epsilon_{2,4}$	4/6+2/6=1	$\epsilon_{5,l}$	2/5	$\epsilon_{10,11}$	2/3+2/3=4/3
$\epsilon_{2,5}$	4/6+2/5=16/15	$\epsilon_{6,11}$	4/6+2/3=4/3	$\epsilon_{10,l}$	2/3
$\epsilon_{2,g}$	4/6=2/3	$\epsilon_{6,12}$	4/6+2/3=4/3	$\epsilon_{11,l}$	2/3
$\epsilon_{2,l}$	4/6=2/3	$\epsilon_{6,13}$	4/6+2/4=7/6	$\epsilon_{12,13}$	2/3+2/4=7/6
$\epsilon_{3,4}$	4/4+2/6=4/3	$\epsilon_{6,g}$	4/6=2/3	$\epsilon_{12,l}$	2/3
$\epsilon_{3,g}$	4/4=1	$\epsilon_{6,l}$	4/6=2/3	$\epsilon_{13,14}$	2/4+2/3=7/6
$\epsilon_{3,l}$	4/4=1	$\epsilon_{7,8}$	2/3+3/2=13/6	$\epsilon_{13,l}$	2/4=1/2
$\epsilon_{4,5}$	2/6+2/5=11/15	$\epsilon_{7,9}$	2/3+2/5=16/15	$\epsilon_{14,l}$	2/3

Algorithm 3.2 Unreliability judgment scheme

Input: The set of compromised line meters $\mathcal{L}$; set of compromised breaker monitors $\mathcal{B}$; grid topology $\mathcal{G}$; the MxST $\mathcal{S}$ of $\mathcal{G}$; number of unrecovered detected bad line meters N_L and circuit breaker monitors N_B

Output: The decision outcome O

1: *Initialization*: the threshold N_{th} denoting the maximum number of unrecovered detected bad devices that a power system can tolerate before a system malfunction
2: **if** $N_L > N_{th}$ or $N_B > N_{th}$ **then**
3:     $O \leftarrow$ system malfunction
4: **else if** at least one pair of compromised line meters and breaker monitors are deployed at a same branch **then**
5:     The attacker is capable of launching a topology attack.
6:     **if** at least one of the compromised line meters and breaker monitors are *critical devices* **then**
7:         $O \leftarrow$ system failure
8:     **else**
9:         $O \leftarrow$ system disturbance
10:     **end if**
11: **else**
12:     $O \leftarrow$ bad data detected with no system unreliability
13: **end if**
        return O

3.4.2.2   Unreliability Judgment Scheme

MxST is employed in our analytical model with the expectation hoping to look for the most critical branches, which provide the most useful information of power system operation status. Thus, the *critical devices* and *critical data* can be defined properly.

Definition 3.1 Given a power grid topology $\mathcal{G}$'s MxST $\mathcal{S}$, sensing devices are termed as *critical devices* if they host on branches ϵ_{ij} which are involved in $\mathcal{S}$, that is

$$\epsilon_{ij} \in \mathcal{G} \cap \mathcal{S}, \ i, j \in \{1, 2, \cdots, N_S\} \text{ and } i \neq j, \tag{3.15}$$

where N_S is the number of edges in $\mathcal{S}$; otherwise, they are termed as *non-critical devices* if

$$\epsilon_{ij} \in \mathcal{G} \setminus \mathcal{S}, \ i, j \in \{1, 2, \cdots, N_S\} \text{ and } i \neq j. \tag{3.16}$$

Hence, the status data generated and the measurement data obtained by these *critical devices* are *critical data*, likewise, the data are termed as *non-critical data* if they are generated by the *non-critical devices*.

An unreliability judgment scheme is designed using the above definitions and the constructed MxST to illustrate requirements which can fire transitions T_DIST, T_FAIL, and T_MALF to cause system unreliability. It can be obtained from Algorithm 3.2 that checking whether the compromised devices are capable of co-constructing a topology attack should be the first step. From the attackers' perspective, a most optimistic condition is considered that an adversary is considered being able to launch a topology attack if the line meter and the circuit breaker monitor on the same branch are unfortunately compromised by the adversary and undetected by the IDS. Otherwise, it is easy to use the bad data detector to detect the injected false meter data. After that, at the moment when an attacker has the ability to launch a topology attack, we further classify to determine whether a system failure or interference has occurred. It should be noted that this classification procedure is also largely successful in distinguishing conservative topology attacks from aggressive topology attacks. It is because, concerning to the definition provided in Sect. 3.3.2, conservative topology attacks usually result in system disturbances, while attacks of aggressive topology usually result in system failures. The time complexity of Algorithm 3.2, which is mainly determined by the operation as Line 4 shows, is $O(N_L \times N_B)$.

3.5   Performance Evaluation

3.5.1   Performance Metrics

In this chapter, the method of combining transient analysis with steady-state analysis is used to analyze the reliability performance of smart grid.

3.5.1.1   Transient Analysis

Mean disturbance time (MTTD) represents the average time before disturbance of the power system, while mean failure time (MTTE) represents the average time before failure. They are indicators of transient analysis. They are given by

$$\text{MTTD} = \int_0^\infty t[1 - Q_D(t)]dt, \tag{3.17}$$

and

$$\text{MTTF} = \int_0^\infty t[1 - Q_F(t)]dt, \tag{3.18}$$

where the first visit's possibility to a system disturbance is denoted by $Q_D(t)$, and the possibility of the first visit to a current system failure is denoted by $Q_F(t)$. It should be noted that in transient analysis, compared with MTTD and MTTF, the mean time to malfunction (MTTM) is ignored because the probability of system malfunction is negligible and MTTM is much smaller.

3.5.1.2   Steady-State Analysis

A self-defined metric: *reliability* R denotes the steady-state analysis, which is defined by

$$R = (1 - p_{malf}) * (1 - \frac{\alpha * p_{dist} + \beta * p_{fail}}{\alpha + \beta})^k, \tag{3.19}$$

where

$$p_{malf} = 1 - \sum_{i=0}^{N_{th}} \sum_{j=0}^{N_{th}} p_{dblm}(i) p_{dbbm}(j), \tag{3.20}$$

denoting if the quantities of detected bad devices which are unrecovered in either place P_DBLM or P_DBBM exceed the threshold N_{th} that can be accepted, a system malfunction occurrence's steady-state probability. The probability of stability about system disturbance and failure is represented by p_{dist} and p_{fail}, respectively. In addition, the negative impacts of p_{dist} and p_{fail} posed to the system reliability are denoted α and β. Under the optimistic condition, the pairs of compromised line meters and breaker monitors hosted at the same transmission line's average number are denoted by k, describing attack events' average number in the power system. In practice, the power system usually has a sufficient recovery rate, wherein a really small value can usually be satisfied. Thus, $p_{dblm}(i)$ and $p_{dbbm}(j)$ always have high probabilities when i and j are small values (e.g., 0 or 1).

Fig. 3.12 The workflow of calculating reliability

According to Eq. (3.20), at steady state, p_{malf} approaches to negligible zero. In this case, Eq. (3.19) can be descended to

$$R = \left(1 - \frac{\alpha * p_{dist} + \beta * p_{fail}}{\alpha + \beta}\right)^k. \tag{3.21}$$

Since absorbing states are P_DIST and P_FAIL's steady states, we are facing difficulties to find the corresponding steady-state probabilities. Hence, this problem is transformed into several small problems. Figure 3.12 shows a corresponding workflow. By temporarily removing places P_DIST and P_FAIL (P_MALF as well), along with the corresponding transitions T_DIST and T_FAIL (T_MALF as well), we reduce the SPN model in this workflow. Next, with the use of the remaining SPN's steady-state analysis, we can easily obtain the steady-state probabilities of $p_{ublm}(i)$ and $p_{ubbm}(j)$, where $i, j \in \{1, 2, \cdots, N\}$. Then, the promising possibility of constructing a topology attack p_{att} is determined, which is given by

$$p_{att} = \sum_{i=1}^{N}\sum_{j=1}^{N} p_{ublm}(i) p_{ubbm}(j) \times \begin{cases} \dfrac{\sum_{l=1}^{\bar{i}} C_{l}^{\bar{i}} C_{\bar{j}-l}^{N-\bar{i}}}{C_{\bar{j}}^{N}}, & i+j \le N \\ 1, & i+j > N, \end{cases} \tag{3.22}$$

where $\bar{j} = \max\{i, j\}$ and $\bar{i} = \min\{i, j\}$. The l (out of $\bar{i}$) pairs' probability of breaker monitors and undetected compromised line meters with a same host location is calculated by $\dfrac{C_{l}^{\bar{i}} C_{\bar{j}-l}^{N-\bar{i}}}{C_{\bar{j}}^{N}}$.

The average probability of at least one pair of undetected compromised line meters and breaker monitors with a same host location is calculated by the dual summations, which is the optimistic condition described in the above section. Then, the steady-state probabilities p_{dist} and p_{fail} can be given by

$$p_{dist} = p_{att} \frac{\overline{N}_S}{N}, \tag{3.23}$$

and

$$p_{fail} = p_{att} \frac{N_S}{N}, \tag{3.24}$$

where $\overline{N}_S$ and N_S are the number of non-spanning tree branches and spanning tree branches, respectively. In particular, compromised bad devices are always sufficient in places P_UBLM and P_UBBM when the compromising rate is large enough. Hence, $i + j > N$ in Eq. (3.22) is always satisfied so that we always have $p_{att} = 1$. Thus, the reliability R is reduced as

$$R = \left(1 - \frac{\alpha * p_{att} * \overline{N}_S/N + \beta * p_{att} * N_S/N}{\alpha + \beta}\right)^k = \left(1 - \frac{\alpha * \overline{N}_S/N + \beta * N_S/N}{\alpha + \beta}\right)^k. \tag{3.25}$$

It should be noted that the attack events' average number in the power system is denoted by k, which is given by

$$k = \sum_{x=1}^{N} x * p(x), \tag{3.26}$$

where

$$p(x) = \sum_{i=x}^{N} \sum_{j=x}^{N} p_{ublm}(i) p_{ubbm}(j) \times \begin{cases} \dfrac{C_x^i C_{j-x}^{N-i}}{C_j^N}, & i + j \leq N \\ 1, & i + j > N \end{cases} \tag{3.27}$$

calculate the x pairs' probability of undetected compromised line meters and breaker monitors which are in an identical host . After that, the expected value of the probability distribution determines k.

3.5.2   Numerical Results

With respect to transient and steady-state analysis, we use PIPE 2 [43] and MATLAB 2015a as our simulators in the simulation experiments, respectively. We set the compromise rate, detection rate, and recovery rate at the same level for the two sensing devices, namely: $\lambda_{clm} = \lambda_{cbm}$, $\lambda_{dblm} = \lambda_{dbbm}$, and $\lambda_{rlm} = \lambda_{rbm}$ to facilitate comparison. The IDS detection interval is set to 10 hours. It is important to note that in this section, since different damage levels simulate virtually all levels of attack capability, conservative topology attacks and aggressive topology attacks are not clearly distinguished. The following figures show the results of all simulations.

The MTTD and MTTF of the IEEE 14-bus test system versus the compromising rate $\lambda_{clm} = \lambda_{cbm}$ for different detection rates $\lambda_{dblm} = \lambda_{dbbm}$ are plotted in Fig. 3.13. It should be noted that, while analyzing the compromising rate and the detection rate, an ample recovery rate of $\lambda_{rlm} = \lambda_{rbm} = 0.08$ is used as a constant parameter. The conclusion can be drawn from Fig. 3.13 that the power system has good operating conditions, and both levels of MTTD and MTTF are significantly high

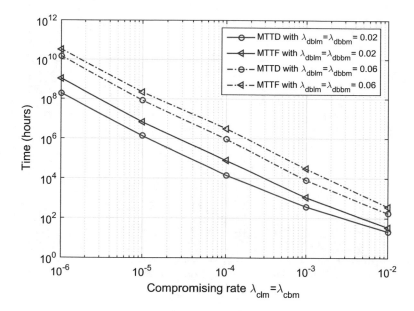

Fig. 3.13 The MTTD and MTTF vs. the compromising rate for IEEE 14-bus system ($\lambda_{rlm} = 0.08$)

when the compromising rate is relatively small. Lower MTTD and MTTF levels can be caused by larger compromising rates as this means more sensing devices are exposed to the network, increasing the likelihood of an adversary launching a topology attack. Furthermore, it can be observed that MTTF is usually larger than MTTD. The reason is that the adversary is more likely to construct a relatively weak attack which can cause system disturbance when our knowledge and capability are limited, rather than a complex attack which can result in system failures. In addition, improving the detection rate can also increase the levels of MTTD and MTTF, for example, from 0.02 to 0.06.

The MTTD and MTTF of the IEEE 14-bus test system versus the detection rate $\lambda_{dblm} = \lambda_{dbbm}$ for different compromising rates $\lambda_{clm} = \lambda_{cbm}$ are shown in Fig. 3.14. It is much more possible to have the compromised bad sensing devices detected and prevent them from initiating topology attacks with high detection rates, which leads to higher levels of MTTD and MTTF. Therefore, it is clear that MTTD and MTTF ascend rapidly as the detection rate increases. From Fig. 3.14, we can obtain similar observation that the MTTD and MTTF levels can be contributed by lower compromising rates as well.

Compared to IEEE 14-bus test system, we also carry out some analogous cases pertaining to the MTTD and MTTF in IEEE 24-bus and 39-bus test systems. We plot the conclusion in Figs. 3.15 and 3.16 for the compromising rate and detection rate, respectively. As Fig. 3.15 shows, analogous to all the three test systems, the larger the compromising rates, the lower the levels of MTTD and MTTF, while the smaller the compromising rates, the higher the levels of MTTD and MTTF.

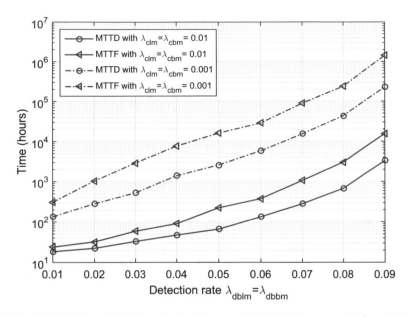

Fig. 3.14 The MTTD and MTTF vs. the detection rate for IEEE 14-bus system ($\lambda_{rlm} = 0.08$)

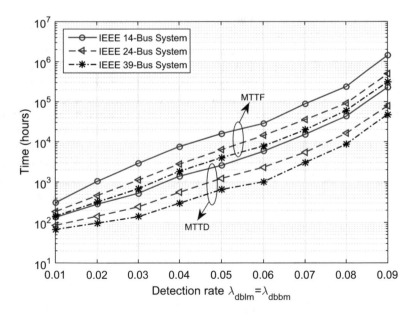

Fig. 3.15 The MTTD and MTTF vs. the compromising rate for various IEEE testing power systems ($\lambda_{dblm} = 0.02$ and $\lambda_{rlm} = 0.08$)

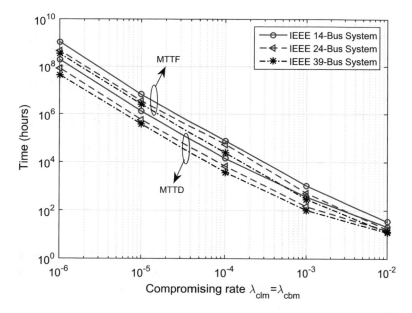

Fig. 3.16 The MTTD and MTTF vs. the detection rate for various IEEE testing power systems ($\lambda_{clm} = 0.001$ and $\lambda_{rlm} = 0.08$)

Most importantly, under the same level of compromising rate, detection rate, and the recovery rate, the IEEE 14-bus system has the highest levels of MTTD and MTTF, followed by the IEEE 24-bus system, then the IEEE 39-bus system. As the total number of sensing devices increases, the average number of compromised sensors per unit time also increases. In this case, it is easier to launch topology attacks, resulting in relatively low MTTD and MTTF levels. It can be seen from Fig. 3.16 that parallel results show that MTTD and MTTF for all three test systems grow exponentially as the system detection rate increases, and IEEE 14-bus system has the highest levels of MTTD and MTTF, while the IEEE 39-bus system has the lowest.

Now we present the steady-state analysis after transient analysis' numerical results are presented. Using the PIPE 2 simulator, we can get the steady-state probability distribution of the number of tokens per location. The steady-state probability distribution of the number of tokens in place P_UBLM (also P_UBBM) under various compromising rates and detection rates is plotted in Figs. 3.17 and 3.18, respectively, for IEEE 14-bus system. As Fig. 3.17 presents, when the compromising rate is relatively low, there is only a few good devices which may be transferred to bad ones and mostly, the devices still stay in good condition, $\Pr\{\#P\_UBLM = 0\}$ is significantly higher, and the probability of other number of tokens is very small. By improving the compromising rate, we can increase $\Pr\{\#P\_UBLM > 0\}$. Contrarily, it can be observed from Fig. 3.18 that a smaller detection rate value of, for example, $\lambda_{dblm} = \lambda_{dbbm} = 0.015$ ($\lambda_{dblm} = \lambda_{dbbm} = 0.04$ in Fig. 3.17), will cause more undetected bad compromised devices, which

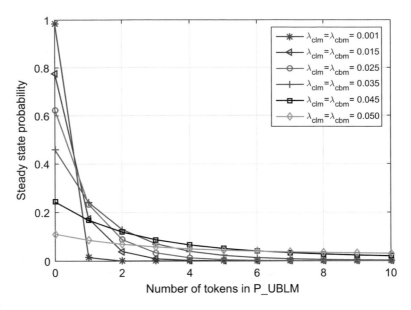

Fig. 3.17 Steady-state probability distribution of the number of tokens in place P_UBLM under different compromising rates for IEEE 14-bus system ($\lambda_{dblm} = 0.04$ and $\lambda_{rlm} = 0.08$)

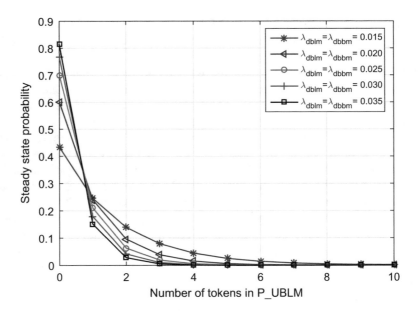

Fig. 3.18 Steady-state probability distribution of the number of tokens in place P_UBLM under different detection rates for IEEE 14-bus system ($\lambda_{clm} = 0.01$ and $\lambda_{rlm} = 0.08$)

cause relatively larger Pr{#*P_UBLM* > 0}. Pr{#*P_UBLM* = 0} can be improved by increasing the detection rate; therefore, it will reduce the potential for an adversary to launch a topology attack.

Using steady-state probability distribution which is obtained previously, Eq. (3.21) can determine the *reliability* of system. Figure 3.19 shows the system reliability of IEEE 14-bus system versus the compromising rate for various values of α/β. The α to β ratio is used in our experiments since based on Eq. (3.21), the system reliability's definition R can be written as $R = (1 - \frac{\alpha * p_{dist} + \beta * p_{fail}}{\alpha + \beta})^k = (1 - \frac{p_{dist}}{1 + \beta/\alpha} - \frac{p_{fail}}{\alpha/\beta + 1})^k$. Hence, using the ratio α/β is more convenient for analyzing the system reliability. Figure. 3.19 presents that under the same level of compromising rate, detection rate, and recovery rate, higher values of α/β result in higher reliability. This is because, according to Eqs. (3.23) and (3.24), p_{fail} is usually greater than p_{dist} due to $N_S > \overline{N}_S$. Therefore, add α/β to assign more weight to p_{dist} and less weight to p_{fail}, thus decreasing the value of $\frac{\alpha * p_{dist} + \beta * p_{fail}}{\alpha + \beta}$. Thus, there will be an increase of the resulting value of reliability and vice versa. What is more interesting is that the reliability decreases gradually, from the origin of $R = 1$, rapidly dropping to almost zero as the increase of the compromising rate. The reason is that it is usually possible to detect defective equipment when the compromising rate is less than the detection rate, thus achieving high reliability, but when the compromising rate is large enough to exceed the detection rate, there will be a large number of unsuccessfully detected defective equipment. In this case, sufficient bad devices together can be easily found in places P_UBLM and P_UBBM that can easily initiate topology attacks, i.e., $p_{att} = 1$ holds all the time. What is more, when the compromising rate is higher, the k will be larger, which indicates the presence of multiple topology attacks and the large value of k will cause the reliability's decrease in an exponential manner.

As shown in Fig. 3.20, it is obvious that there is a linear relationship between the system reliability of IEEE 14 bus system and the detection rate of α/β. Similarly, it can be concluded from this figure that under the same conditions, the larger the value of α/β, the higher the system reliability, while smaller value of α/β can lead to lower system reliability. Furthermore, it can be seen from the full simulation trace that at the beginning, the system reliability experiences a slight growth, but with the increase of detection rate, it eventually reaches a plateau at around $R = 1$. This indicates that the number of undetected compromised devices can be slowly decreased with the improvement of the detection rate, then the probability of initiating an attack will be reduced as well, which further increases the system reliability.

In addition, we also conduct simulation experiments concerning steady-state analysis for the IEEE 24-bus and 39-bus systems. Figure 3.21 shows the system reliability of the three test systems against the compromising rate under various values of α/β. We can obtain similar results to the IEEE 14-bus system for the IEEE 24-bus and 39-bus systems. Specifically, for all the three test systems, as the compromising rate increases, the reliability decreases gradually from $R = 1$, and when the compromising rate exceeds the detection rate, it drops quickly.

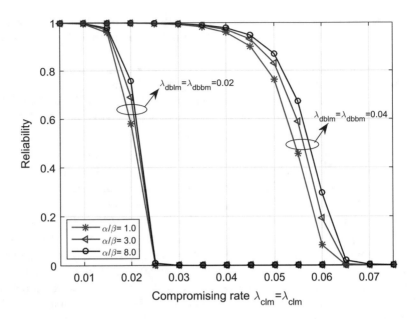

Fig. 3.19 System reliability of IEEE 14-bus system vs. the compromising rate ($\lambda_{rlm} = 0.08$)

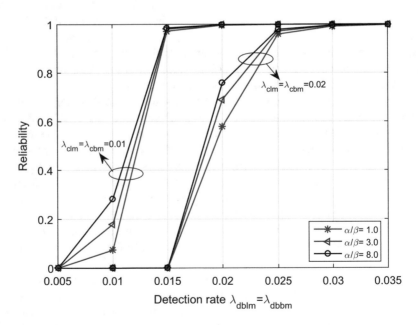

Fig. 3.20 System reliability of IEEE 14-bus system vs. the detection rate ($\lambda_{rlm} = 0.08$)

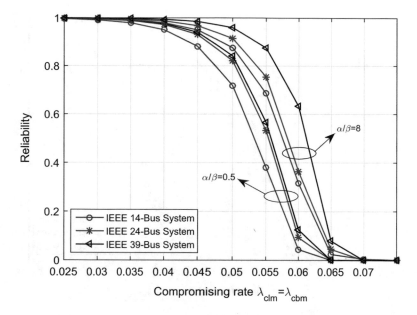

Fig. 3.21 System reliability vs. the compromising rate for various IEEE testing power systems ($\lambda_{dblm} = 0.04$ and $\lambda_{rlm} = 0.08$)

Furthermore, it can be also observed that when α/β is set to be high, the power systems can have a high system reliability. In addition, IEEE 39 bus system has the highest reliability, followed by 24 bus system and 14 bus system, which can be obtained by numerical results. This is because that a power system may be more resilient to the attacks if it has more redundant branches and higher connection complexity.

In Fig. 3.22, the reliability numerical diagrams of different power systems are drawn according to the different detection rates. Similarly, the reliability stabilizes at nearly zero at the original period in each test system, but a slight difference is that the stabilization results from insufficiency of the detection rate to identify the compromised bad devices. As the detection rate gradually increased, reliability began to increase dramatically, eventually reaching $R = 1$ when the detection rate was sufficient.

In the end, our simulation experiments focus on steady-state analysis against the recovery rate. Figures 3.23 and 3.24 summarize the corresponding numerical results. It can be seen from Fig. 3.23 that under the same compromising rate and detection rate but different recovery rates, the number of tokens' steady-state probability distribution in place P_UBLM is the same, which can be concluded from the three groups of comparative experiments. The reason is that the number of detected bad devices arriving in place P_UBLM is determined by the compromising rate and detection rate collectively; hence, the same distribution of tokens in steady states will be caused by the same compromising rate and detection rate. At the same time, we can conclude that the recovery rate is of little impact on this distribution,

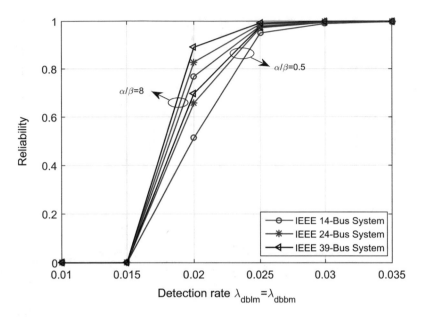

Fig. 3.22 System reliability vs. the compromising rate for various IEEE testing power systems ($\lambda_{clm} = 0.02$ and $\lambda_{rlm} = 0.08$)

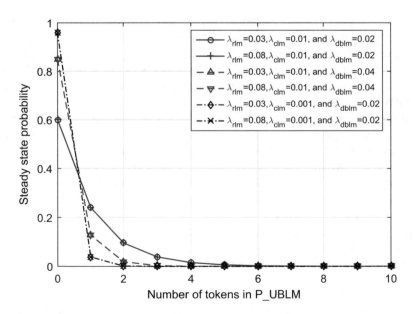

Fig. 3.23 Steady-state probability distribution of the number of tokens in place P_UBLM under different recovery rates for IEEE 14-bus system

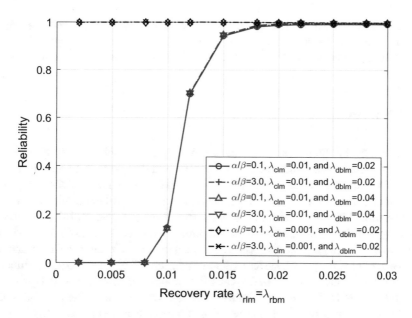

Fig. 3.24 System reliability vs. the recovery rate for IEEE 14-bus system

i.e., there is no significant difference between the recovery rate and the probability of topology attacks. The other curves once again show that improving the reducing rate or detecting the compromising rate can further mitigate the number of compromised devices in place P_UBLM.

As shown in Fig. 3.24, it can be observed that, as the recovery rate from the two groups of red and blue curves increases slightly, the system reliability increases dramatically from 0 to 1. In such circumstances, there are system malfunctions if the recovery rate is not sufficient (e.g., $\lambda_{rlm} = \lambda_{rbm} < 0.02$ here) to have the detected bad devices which under a compromising rate of $\lambda_{clm} = \lambda_{cbm} = 0.01$ and a detection rate of either $\lambda_{dblm} = \lambda_{dbbm} = 0.02$ or 0.04 are recovered. In contrast, as the two black curves show that under a rather small compromising rate (i.e., $\lambda_{clm} = \lambda_{cbm} = 0.001$), a recovery rate of $\lambda_{rlm} = \lambda_{rbm} = 0.002$ is reasonably sufficient to recover all the detected bad devices. This leads to a full system reliability (i.e., $R = 1$) with no system malfunctions. A sufficient recovery rate of $\lambda_{rlm} = \lambda_{rbm} = 0.08$ is used to eliminate the possible impacts that insufficient recovery rate may cause on system reliability in previous simulations related to the compromising rate and detection rate. In comparison to the compromising rate and detection rate, we can conclude that the system reliability is less likely to be impacted by recovery rate as long as it can reach a basic acceptable level in this figure. What we should pay more attention to is that the compromising rate is highly determined by the recovery rate, while the detection rate has less impact in comparison.

Such observations will help guide future system design and enable system designers to work harder on important aspects, such as reducing compromising rate and increasing the detection rates, instead of focusing too much on recovery rates.

3.6 Summary

In the foreseeable future, more and more smart grid cyber-physical systems will be deployed, which means more and more research challenges need to be solved.

In this chapter, in order to have the system reliability in the presence of topology attack and system game evaluated and analyzed, a smart grid cyber-physical system analysis model based on SPN is developed. In addition, we successfully constructed topology attacks in a smart grid and demonstrated the process. We considered two types of sensing devices (i.e., line meters and circuit breaker monitors) and two kinds of typical system countermeasures (i.e., IDSs and malfunction recovery techniques) in our analytical SPN model and demonstrated how to use multiple events to describe the behavior of the system when these events occur. In addition, taking IEEE 14 bus as an example, we propose two algorithms to construct MxST and identify system interference and fault. Moreover, simulation experiments are carried out on the TEST system of IEEE 14 bus, 24 bus, and 39 bus, and corresponding transient and steady-state analyses are obtained, which verifies the practicability and effectiveness of the model. This (e.g., confirming that compromising rate and detection rate are critical to system reliability) will provide reference for future system design.

References

1. Li, B., Lu, R., Wang, W., & Choo, K.-K. R. (Nov. 2016). DDOA: A Dirichlet-based detection scheme for opportunistic attacks in smart grid cyber-physical system. *IEEE Transactions on Information Forensics and Security, 11*(11), 2415–2425.
2. Choo, K.-K. R. (Apr. 2014). A conceptual interdisciplinary plug-and-play cyber security framework. In *ICTs and the Millennium Development Goals* (pp. 81–99). New York: Springer Science & Business Media.
3. Li, B., Lu, R., Wang, W., & Choo, K.-K. R. (May 2017). Distributed host-based collaborative detection for false data injection attacks in smart grid cyber-physical system. *Journal of Parallel and Distributed Computing, 103*, 32–41.
4. Moslehi, K., & Kumar, R. (June 2010). A reliability perspective of the smart grid. *IEEE Transactions on Smart Grid, 1*(1), 57–64.
5. Li, B., Lu, R., & Bao, H. (2016). Behavior rule specification-based false data injection detection technique for smart grid. In *Cyber Security for Industrial Control Systems: From the Viewpoint of Close-Loop* (pp. 119–150).
6. Liu, Z., Zhang, C., Dong, M., Gu, B., Ji, Y., & Tanaka, Y. (2016). Markov-decision-process-assisted consumer scheduling in a networked smart grid. *IEEE Access, 5*, 2448–2458.
7. Farhangi, H. (Jan.-Feb. 2010). The path of the smart grid. *IEEE Power and Energy Magazine., 8*(1), 18–28.

8. Amin, S. M., & Wollenberg, B. F. (Sep.-Oct. 2005). Toward a smart grid: power delivery for the 21st century. *IEEE Power and Energy Magazine, 3*(5), 34–41.
9. U. F. E. R. C. (FERC). (2009). Smart grid policy. *Docket PL09-4-000.*
10. Deng, R., Xiao, G., Lu, R., Liang, H., & Vasilakos, A. V. (Apr. 2017). False data injection on state estimation in power systems—attacks, impacts, and defense: A survey. *IEEE Transactions on Industrial Informatics, 13*(2), 411–423.
11. Wu, J., Dong, M., Ota, K., Zhou, Z., & Duan, B. (June 2014). Towards fault-tolerant fine-grained data access control for smart grid. *Wireless Personal Communications, 75*(3), 1787–1808.
12. Li, W. (Mar. 2014). *Risk assessment of power systems: models, methods, and applications.* New York: John Wiley & Sons.
13. Falliere, N., Murchu, L. O., & Chien, E. (Feb. 2011). W32. Stuxnet dossier. *White Paper, Symantec Corp., Security Response, 5*(6), 29.
14. Pagliery, J. (Oct. 2015). ISIS is attacking the U.S. energy grid. http://money.cnn.com/2015/10/15/technology/isis-energy-grid/ (Accessed: 2018-06-28).
15. Kröger, W. (Dec. 2008). Critical infrastructures at risk: A need for a new conceptual approach and extended analytical tools. *Reliability Engineering & System Safety, 93*(12), 1781–1787.
16. Lee, E. A. (Feb. 2017). Fundamental limits of cyber-physical systems modeling. *ACM Transactions on Cyber-Physical Systems, 1*(1), 3–26.
17. Kim, J., & Tong, L. (July 2013). On topology attack of a smart grid: undetectable attacks and countermeasures. *IEEE Journal on Selected Areas in Communications, 31*(7), 1294–1305.
18. Dalton, G., Mills, R. F., Colombi, J. M., & Raines, R. A. (2006). Analyzing attack trees using generalized stochastic Petri nets. In *Proc. IEEE Information Assurance Workshop (IAW'06), West Point, NY, USA, June 21–23*, pp. 116–123.
19. Mitchell, R., & Chen, I.-R. (Mar. 2016). Modeling and analysis of attacks and counter defense mechanisms for cyber physical systems. *IEEE Transactions on Reliability, 65*(1), 350–358.
20. Liu, Y., Ning, P., & Reiter, M. K. (May 2011). False data injection attacks against state estimation in electric power grids. *ACM Transactions on Information and System Security, 14*(1), 13.
21. McDonald, R., Pereira, F., Ribarov, K., & Hajič, J. (2005). Non-projective dependency parsing using spanning tree algorithms. In *Proc. Conference on Human Language Technology and Empirical Methods in Natural Language Processing (HLT/EMNLP'05), Vancouver, British Columbia, Canada, Oct. 06–08*, pp. 523–530.
22. Jensen, K., & Rozenberg, G. (Dec. 2012). *High-level Petri nets: theory and application.* New York: Springer Science & Business Media.
23. McDermott, J. P. (2001). Attack net penetration testing. In *Proc. New Security Paradigms Workshop (NSPW'00), Ballycotton, County Cork, Ireland, Sep. 18–21*, pp. 15–21.
24. Bause, F., & Kritzinger, P. (Jan. 1996). *Stochastic Petri Nets*, vol. 26.
25. Tüysüz, F., & Kahraman, C. (May 2010). Modeling a flexible manufacturing cell using stochastic Petri nets with fuzzy parameters. *Expert Systems with Applications, 37*(5), 3910–3920.
26. Jensen, K. (Apr. 2013). *Coloured Petri nets: basic concepts, analysis methods and practical use* (vol. 1). New York: Springer Science & Business Media.
27. Laprie, J.-C., Kanoun, K., & Kaâniche, M. (2007). Modelling interdependencies between the electricity and information infrastructures. In *Proc. International Conference on Computer Safety, Reliability, and Security (SAFECOMP), Nuremberg, Germany, Sep. 18–21*, pp. 54–67.
28. Zeng, R., Jiang, Y., Lin, C., & Shen, X. (Sep. 2012). Dependability analysis of control center networks in smart grid using stochastic Petri nets. *IEEE Transactions on Parallel and Distributed Systems, 23*(9), 1721–1730.
29. Chen, T. M., Sanchez-Aarnoutse, J. C., & Buford, J. (Dec. 2011). Petri net modeling of cyber-physical attacks on smart grid. *IEEE Transactions on Smart Grid, 2*(4), 741–749.
30. Householder, A., Houle, K., & Dougherty, C. (Apr. 2002). Computer attack trends challenge Internet security. *Computer, 35*(4), sulp5–sulp7.

31. Weimer, J., Kar, S., & Johansson, K. H. (2012). Distributed detection and isolation of topology attacks in power networks. In *Proc. 1st International Conference on High Confidence Networked Systems (HiCoNS), Beijing, China, Apr. 17–18*, Beijing, China, pp. 65–72.
32. Liu, S., Liu, X. P., & El Saddik, A. (2013). Denial-of-Service (DoS) attacks on load frequency control in smart grids. In *Proc. Innovative Smart Grid Technologies (ISGT), Washington, DC, USA, Feb. 24–27*, pp. 1–6.
33. Amin, S., Cárdenas, A. A., & Sastry, S. S. (2009). Safe and secure networked control systems under Denial-of-Service attacks. In *Proc. International Workshop on Hybrid Systems: Computation and Control (HSCC), San Francisco, CA, USA, Apr. 13–15*, pp. 31–45.
34. Hug, G., & Giampapa, J. A. (Sep. 2012). Vulnerability assessment of ac state estimation with respect to false data injection cyber-attacks. *IEEE Transactions on Smart Grid, 3*(3), 1362–1370.
35. Djekic, Z. (Dec. 2007). Online circuit breaker monitoring system. Ph.D. dissertation, Texas A&M University, College Station, TX.
36. Boneh, D., Lynn, B., & Shacham, H. (Sep. 2004). Short signatures from the Weil pairing. *Journal of Cryptology, 17*(4), 297–319.
37. Huang, L., Sun, Y., Xu, J., Gao, W., Zhang, J., & Wu, Z. (Mar. 2014). Optimal PMU placement considering controlled islanding of power system. *IEEE Transactions on Power Systems, 29*(2), 742–755.
38. Vildhøj, H. W., & Wind, D. K. (2016). *Supplementary notes for graph theory* (vol. 1, pp. 26–26). [Online]. Available: http://www.student.dtu.dk//~dawi/01227/01227-GraphTheory.pdf
39. Grigg, C., Wong, P., Albrecht, P., Allan, R., Bhavaraju, M., Billinton, R., et al. (Aug. 1999). The IEEE reliability test system-1996. a report prepared by the reliability test system task force of the application of probability methods subcommittee. *IEEE Transactions on Power Systems, 14*(3), 1010–1020.
40. Hoare, C. A. (1962). Quicksort. *The Computer Journal, 5*(1), 10–16.
41. Graham, R. L., & Hell, P. (Jan.-Mar. 1985). On the history of the minimum spanning tree problem. *Annals of the History of Computing, 7*(1), 43–57.
42. Kruskal, J. B. (Feb. 1956). On the shortest spanning subtree of a graph and the traveling salesman problem. *Proceedings of the American Mathematical Society, 7*(1), 48–50.
43. Dingle, N. J., Knottenbelt, W. J., & Suto, T. (Mar. 2009). PIPE2: A tool for the performance evaluation of generalised stochastic Petri nets. *ACM SIGMETRICS Performance Evaluation Review, 36*(4), 34–39.

Chapter 4
DHCD: Distributed Host-Based Collaborative Detection for FmDI Attacks

The entire power system may be suffer catastrophic damage caused by FmDI attacks. However, since the high reliance on open information networks, it is challenging to counter FmDI attacks in smart grid CPS. Although many solutions for FmDI attacks have been proposed, most of them are based on state estimation with the help of the highly centralized control center; therefore, computationally expensive. Moreover, high-level security assurance generally does not exist in these solutions, as demonstrated by recent event that smart FmDI attackers with knowledge of system configurations can easily bypass false data detection mechanisms based on the conventional state estimation. In this chapter, we present a novel distributed host-based collaborative detection method in order to address these challenges. To be specific, in order to collaboratively detect false data as well as those PMUs with false measurement data, a conjunctive rule based majority voting algorithm is designed in the proposed method. Furthermore, an innovative reputation system with an adaptive reputation updating algorithm is also devised. The overall running status of PMUs can be evaluated by the algorithm above, then, FmDI attacks can be clearly observed. The high effectiveness of our approach is demonstrated by extensive experiments on real-time measurement data collected from the PowerWorld simulator.

4.1 Introduction

In recent years, a lot of high-profile incidents aiming at smart grid or other CPSs, such as Stuxnet [1], Conficker [2], and US drones hack [3], have been reported. The purpose of malicious attackers is to fake measurements data, insert false control commands, delay or drop measurement data, or control commands [4–8]. With the increasing threat to the smart grid CPS, FmDI attacks, naturally, have been received great attention from computer security researchers and industry practitioners. Over

© Springer Nature Switzerland AG 2020
B. Li et al., *Detection of False Data Injection Attacks in Smart Grid Cyber-Physical Systems*, Wireless Networks,
https://doi.org/10.1007/978-3-030-58672-0_4

the years, both FmDI attacks and mitigation strategies on smart grid CPS have been also developed. Conventional FDD approaches are generally based on system state estimation [9–11]. For instance, for the purpose of improving the performance of static state estimation, Merrill and Schweppe proposed a bad data suppression estimator based on a non-quadratic cost function[9]. Handschin et al. proposed a method for the purpose of detecting and identifying the bad data and structural error problems, and improved bad data analysis (detection probability, and effects of bad data) [8]. Cutsem et al. presented a detection method in order to mitigate some existing difficulties, e.g., multiple and interacting bad data [12].

However, Liu et al. in [13] showed that smart FmDI attackers, if know the current system configurations, can easily circumvent the legacy state estimation-based FDD schemes. Hence existing FDD approaches may not be able to effectively counter against newer or emerging FmDI attacks. Traditional FDD methods have many limitations, the major one is that they mainly pay attention to the inter-correlations among the measurement data, such as residuals and errors, rather than the malicious behaviors of meter devices (e.g. smart meters and PMUs). In addition, the existing FDD is usually performed by the centralized CC of the power system, due to its computational complexity [10, 11]. Although the computational burden of the CC can be effectively reduced by a small amount of distributed or hierarchical FDD schemes [14, 15], most of these schemes are still grounded on state estimation; therefore, may be bypassed easily. Another shortcoming of traditional FDD schemes is that some prevalent detection methods only detect the "bad" data, rather than further evaluating the true running status of those compromised devices [8, 9, 14]. Thus, the attacks can be continuously launched or improved by these undetected attackers. Therefore, countering against FmDI attacks in smart grid CPS remains a research challenge, and one that we seek to address in this chapter.

In this chapter, different from legacy methods based on state estimation, we present a distributed host-based collaborative detection (DHCD) method based on rule specifications. The proposed DHCD is lightweight in terms of computation, so it can achieve rapid FDD and the ability to identify compromised devices by assessing the running status of them. To be specific, in the proposed approach, each PMU is assigned a creditable distributed local false data detector, referred to as host monitor (HM) in this chapter. According to a series of pre-defined rule specifications, the anomalous levels of measurement data collected by PMUs are determined by the monitors. After that, the anomalous levels of the measurement data, which are collected by the neighboring interconnected PMUs, will be shared and compared, these interconnected monitors collaboratively then determine whether their own measurement data is illegally modified by using the majority voting algorithm. Moreover, with the adaptive reputation updating (ARU) algorithm, we can assess the running status of PMUs and then easily identify which one is compromised. We regard the contributions of this chapter to be threefold:

- A DHCD method based on rule specifications is proposed for the purpose of detecting and mitigating FmDI attacks in smart grid CPS.

- Our approach can achieve fast and high accuracy of FDD, while also allowing the detection of malfunction PMUs using our designed reputation system.
- The computational burden of the CC will be effectively reduced by our approach since FDD tasks are delegated to the local monitors.

The remainder of this chapter is organized as follows. The system model, the threat model, and our design goals are described in Sect. 4.2. The detail of our proposed DHCD method is presented in Sect. 4.3, then, Sect. 4.4 gives the performance evaluation. Finally, Sect. 4.5 concludes the chapter with future research directions.

4.2   Models and Design Goals

In this section, we present the system model, the threat model, as well as the design goals.

4.2.1   System Model

A smart grid CPS is a fully automatic system with the capacity to achieve self-healing, reduce cost, and improve reliability and efficiency. Such benefits need to be promised by the WAMS, which can provide advanced observability and controllability in the operation of power system [16–18]. Therefore, in this chapter, the hierarchical WAMS is concerned as our system model.

Figure 4.1 shows that WAMS is an integrated system composed of PMUs, PDCs, heterogenous communication networks, and a CC. Particularly, the real-time status of the power system is measured by PMUs deployed in the substations. For instance, the PMUs can measure the real-time amplitude and phase angle of the power at each branch, of current on the transmission line, and of voltage at the bus. These data measured by PMUs are then periodically delivered to the PDCs, usually in 50 Hz, through the LAN. After that, the CC will obtain the aggregated data from the PDCs via the WAN. Finally, the measurements data will be analyzed at the CC (e.g., state estimation, event diagnostics, and contingency analysis).

4.2.2   Threat Model

The real measurement data collected from PMUs can ensure that the system operates efficiently and reliably. Nevertheless, the attacks with inserting false data through malfunction PMUs will destroy the authenticity of the measurement data. Successful FmDI attack may compromise the above-mentioned promising functionalities or even jeopardize the system operations.

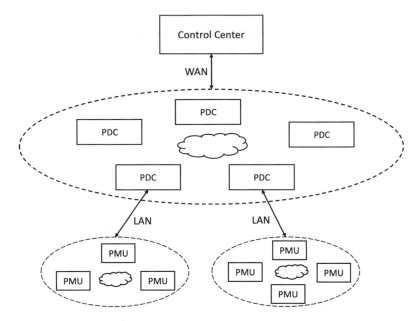

Fig. 4.1 The hierarchical architecture of WAMS

In our threat model. we consider that PMUs in the WAMS can be compromised by FmDI attackers (e.g., rewriting the program settings, or stealing the secret information for data communication). Specifically, in smart grid CPS, since the system can correct trivial faults or mistakes, a single false sensor data may not have great impact on operations of the system. However, if consecutive false measurement data are received, the system may not be able to auto-correct; then, leading to system failures. Therefore, we consider that attackers, if they have an opportunity, usually persistently and recklessly insert false data in order to successfully launch an FmDI attack.

4.2.3   Design Goals

The main design goal in this chapter is to address the above-mentioned threatening issues by devising a scalable, accurate, and efficient FDD method. To be more specific, the following goals should be achieved.

Accuracy   The proposed method should achieve high detection rate as well as low false alarm rate in terms of detecting smart FmDI attacks.

Efficiency   Unlike legacy FDD methods deployed in CC, the detection method should be lightweight in terms of computation, that is, the proposed method will not bring a large amount of calculation burden to the system, especially to the CC.

Scalability Similar to a cloud system, the smart grid CPS should have scalability, that is, new devices are allowed to be deployed without incurring large (financial) costs.

4.3 Proposed DHCD Method

In this section, the detail of our proposed DHCD method including two steps (subsections): collaborative FDD and detection of malfunction PMU is provided. First, the measurement data collected by the PMU is analyzed, then, the anomalous data can be identified by using a line of rule specifications. After that, we design a reputation system with an ARU algorithm to monitor and evaluate PMUs' running status for the purpose of further detection of the PMU with false data.

4.3.1 Collaborative FDD

Under normal conditions, the power grid will operate in a stable state. That is, in light of Kirchhoff's law, demand-response constraints, etc., all state variables change in a mutual balanced manner. As such, any change of a variable state on one bus or transmission line, resulting from either the normal demand variation or system faults, would lead to corresponding state changes of the same and/or other variables on interconnected buses or transmission lines. For instance, Fig. 4.2a, b shows the variation of the current amplitude on each transmission line when an open circuit event occurs on transmission line from Bus 16 to Bus 17. It can be seen from Fig. 4.2b that the current amplitude values near Lines 16–17 shift after such an event occurs. The closer to this line, the more the value changes.

On the contrary, we can regard some changes that only happen on one bus without the corresponding change of variable states on interconnected buses, as abnormal. These abnormalities may be caused by either malicious activities or malfunction devices. Since there are many present methods to solve problems related to device malfunction, only possible malicious activities are considered in this chapter. More specifically, a collaborative detection method, which is grounded on the inter-correlations of power systems, is proposed by us to detect abnormal measurement data collected by PMUs [19, 20].

4.3.1.1 Normal Rule Specifications

If the power system is operating normally, all state variables must naturally follow certain rules. For instance, in an example of active power P, it should obey the rules shown below:

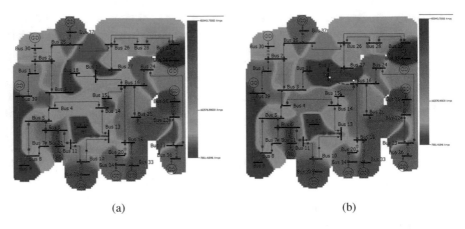

(a)   (b)

Fig. 4.2 Comparison of contouring maps describing the distribution of current amplitude on transmission lines: (**a**) before open circuit and (**b**) after open circuit on line from Bus 16–17 (marked by a red dashed elliptical circle) in IEEE-39 bus system

Table 4.1 Rules specifications for PMUs in stable status	Index	Variable	Rule description
	1	Active power angle	$\Delta\delta < \delta_\Delta$
	2	(Phase A) voltage amplitude	$\Delta V < V_\Delta$
	3	Load Mvar	$\Delta L_{Mvar} < L_{Mvar\Delta}$
	4	Load MW	$\Delta L_{MW} < L_{MW\Delta}$

- $P_{min} < P^t < P_{max}$: P at any time under stable status must change within the expected range $[P_{min}, P_{max}]$.
- $|P^t - P^{t-1}| < P_\Delta$: The variation of P within a time interval should be less than the experienced threshold P_Δ.
- $|P^t_{in} - P^t_{out}| < P_{loss}$: The difference between the P flowing into a bus and flowing out the bus should be less than an experienced power loss threshold P_{loss}.
- Other more complicated rules.

As such, we pre-define some useful but example rule specifications as listed in Table 4.1 that PMUs have to coincide with in the stable status. We regard these rule specifications as the foundation of the proposed approach to detect the abnormal measurement data (the superscript t is omitted for convenience).

In order to indicate the results of whether the rule specifications have been followed, a binary system is employed in our DHCD, where "0" represents that the relevant rule specification is followed by current measurement data of one variable, and "1" denotes a violation. The merged results related to the entire measurement data are represented by a binary sequence with length E, which depends on the number of rule specifications (E is set as 4 in this chapter). For example, "1001" indicates the violations of rules 1 and 4. Similarly, "0000" denotes a non-violation of the conjunctive four rule specifications, which is our *baseline* of PMUs' behaviors.

For the purpose of evaluating what extent each piece of measurement data is abnormal, the *anomalous level* l^t is determined by using a normalized Euclidean distance strategy, which is given below:

$$l^t = D_0(seq^t, seq_0), \tag{4.1}$$

where seq^t is the binary sequence indicating the conjunctive results of measurement data at time t, while $seq_0 = $ "0000" is the baseline. D_0 is the normalized Euclidean distance between sequences seq^t and seq_0. Euclidean distance can be used to compute the difference between two sequences. For instance, the Euclidean distance between sequence "1001" and the *baseline* "0000" is $\sqrt{1^2 + 0 + 0 + 1^2} \approx 1.414$. Then, the *anomalous level* l can be obtained by the normalized distance, i.e., $1.414/\sqrt{1^2 + 1^2 + 1^2 + 1^2} \approx 0.707$.

4.3.1.2   FDD Algorithm with Iterative Majority Voting

As shown in Fig. 4.3, in the distributed host-based collaborative FDD system, the task of each HM is to monitor and assess the behaviors of its administrated PMU. Let $\mathcal{M} = \{M_1, M_2, \ldots, M_N\}$ represent the set of monitors and $\mathcal{U} = \{U_1, U_2, \ldots, U_N\}$ represent the set of PMUs, where N is the total number of HMs or PMUs. The communication among HMs follows the PMUs connection mode, which means each HM only communicates with HMs that their monitored PMUs have interconnection relations. Note that these HMs are trusted entities for monitoring PMUs. It is devised that host monitors and the networks between them are equipped with high-level security mechanisms to ensure their trustworthiness.

As mentioned above, the detection method is constructed by using the inter-correlations among the state variables. The FDD algorithm with iterative majority voting process is outlined in Algorithm 4.1. Specifically, the set $\mathcal{M}$ is initialized as $\mathcal{M} = \{M_1, M_2, \ldots, M_N\}$, and a flag variable $repeat\_flag$ as "0." Here, if $repeat_flag = $ "0," the procedure does not need to be repeated, otherwise the procedure need to be repeated. Then, the conjunctive result R_i^t of present piece of measurement data will be determined by monitor $M_i \in \mathcal{M}$, and the result will be broadcasted to neighboring connected monitors $\mathcal{M}_i = \{M_j | M_j \sim M_i\}$. An example is shown in Fig. 4.4.

After that, the false data detection procedure is launched by M_i. If the bits in the result R_i^t are all "0," then no false data is identified. Otherwise, the number of bit "1" in conjunctive results R_j^t corresponding to monitors connected with M_i needs to be determined by M_i. If more than or equal to half of the connected monitors have a bit "1" at the same position in R_j^t, M_i concludes that U_i has reported a piece of false measurement data; on the contrary, R_i^t is temporarily considered suspicious. All $M_i \in \mathcal{M}$ will conclude the first procedure for the determination of termination criterion. If $repeat_flag == $ "0," this procedure is ended; on the contrary, the procedure needs to be repeated for the purpose of further detection of the false data.

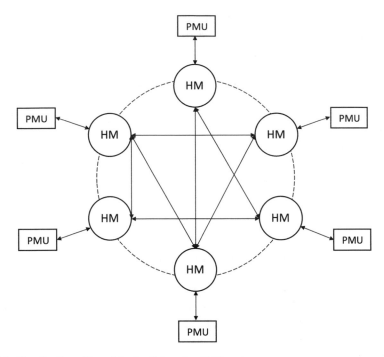

Fig. 4.3 The distributed host-based collaborative FDD system

M_1 :

Rule_Index	R1	R2	R3	R4
Rule_Result	0	0	0	0

M_2 :

Rule_Index	R1	R2	R3	R4
Rule_Result	0	0	0	1

•
•
•

M_N :

Rule_Index	R1	R2	R3	R4
Rule_Result	0	1	0	0

Fig. 4.4 An example of the conjunctive results transmitted between HMs

Here, the time complexity of Algorithm 4.1 is $O(N \times L)$, where $L = \frac{1}{2} \sum_{i=1}^{N} |\mathcal{M}_i|$, the number of branches in a power grid.

Algorithm 4.1 FDD algorithm

 1: **initialization**: $\mathcal{M} = \{M_1, M_2, \ldots, M_N\}$, $Upperbound = 5$, $Iteration = 0$, $repeat\_flag =$ '0'
 2: **procedure**
 3:     **for** each monitor $M_i \in \mathcal{M}$ **do**
 4:         (1). determines the conjunctive result R_i^t of current piece of measurement data.
 5:         (2). broadcasts the result R_i^t to the neighboring connected monitors $\mathcal{M}_i = \{M_j | M_j \sim M_i\}$.
 6:         (3). **identifies false data:**
 7:         **if** there is no bit "1" in the result R_i^t **then**
 8:             **output:** no false data detected.
 9:         **else if** more than or equal to half of the monitors in $\mathcal{M}_i$ hold bit "0" at the same position in the result R_j^t **then**
10:             (a). **output:** false data detected.
11:             (b). removes M_i from $\mathcal{M}$ and its connections with other monitors.
12:         **else**
13:             (a). keeps R_i^t as suspicious result.
14:             (b). $repeat\_flag =$ '1'.
15:         **end if**
16:     **end for**
17:     (4). **judges the termination criteria:**
18:     **if** $repeat\_flag ==$ '1' and $Iteration < Upperbound$ **then**
19:         (a). repeats **procedure**.
20:         (b). $Iteration = Iteration + 1$.
21:     **else**
22:         ends the procedure.
23:     **end if**
24: **end procedure**

4.3.2   Determination of Compromised PMU

FDD step, although it can effectively identify false data, is not enough to identify compromised PMUs. Thence, in the next step, a reputation-based algorithm is adopted to monitor and evaluate the PMUs' overall behaviors over a period of time, with the help of such algorithm, we can determine a compromised PMU when its reputation level falls below an acceptable threshold [21, 22].

Specifically, the probability distribution of the abnormal level of data collected by PMUs is modeled with a Beta distribution. After that, the maximum likelihood estimation (MLE) as well as Newton–Raphson method is used for the purpose of estimating two shape parameters α and β. Thereafter we describe the detail of the ARU algorithm.

4.3.2.1 Probability Distribution of Anomalous Level

The normalized Euclidean distance (see Sect. 4.3.1.1) determines the abnormal level of measurement data, which is represented by random variable X. Here, $X = 0$ denotes compliance with the rule specifications, while $X = 1$ denotes a violation. Note that, in order to determine the exact distribution of the probabilities of different anomalous level and its future values, a $Beta(\alpha, \beta)$ distribution is used to model the random variable X. Beta distribution family can indicate a set of probability distributions, and can be utilized to depict a prior distribution of an unknown distribution with only a line of collected observations.

The probability density function (PDF) of a Beta distribution is

$$f(x; \alpha, \beta) = \frac{\Gamma(\alpha + \beta)}{\Gamma(\alpha)\Gamma(\beta)} x^{\alpha-1}(1 - x)^{\beta-1}, \tag{4.2}$$

where α and β are the two shape parameters. The mean value of a Beta distribution can be given by

$$\mu = E[X] = \int_0^1 x \frac{\Gamma(\alpha + \beta)}{\Gamma(\alpha)\Gamma(\beta)} x^{\alpha-1}(1 - x)^{\beta-1} dx = \frac{\alpha}{\alpha + \beta}. \tag{4.3}$$

A well-known method MLE is utilized to estimate the parameters α and β for the purpose of obtaining exact distribution of X. Specifically, we suppose that the n independent and identically distributed observations $\{x_1, x_2, \ldots, x_n\}$ are from an unknown distribution with PDF $f_0(\cdot|\boldsymbol{\theta})$, $where \boldsymbol{\theta}$ is a vector of parameters. As for our model, the Beta distribution, $\boldsymbol{\theta} = [\alpha\ \beta]$. By using MLE, the joint density probability function of these n independent and identically distributed observations $\{x_1, x_2, \ldots, x_n\}$ is formulated as

$$f(x_1, x_2, \ldots, x_n \mid \alpha, \beta) = \prod_{i=1}^n f(x_i \mid \alpha, \beta). \tag{4.4}$$

Now we consider another scenario that $\{x_1, x_2, \ldots, x_n\}$ are observable samples, then α, β are the variables of the function called the likelihood:

$$\mathcal{L}(\alpha, \beta \mid x_1, x_2, \ldots, x_n) = \prod_{i=1}^n f(x_i \mid \alpha, \beta). \tag{4.5}$$

For the convenience of calculation, the natural logarithm of the likelihood function is adopted, then, the above equation is rewritten as

$$\ln \mathcal{L}(\alpha, \beta \mid x_1, x_2, \ldots, x_n) = \ln \prod_{i=1}^{n} f(x_i \mid \alpha, \beta)$$

$$= \sum_{i=1}^{n} \ln \left\{ \frac{\Gamma(\alpha + \beta)}{\Gamma(\alpha)\Gamma(\beta)} \, x_i^{\alpha-1} (1 - x_i)^{\beta-1} \right\}$$

$$= n \ln \Gamma(\alpha + \beta) - n[\ln \Gamma(\alpha) + \ln \Gamma(\beta)]$$

$$+ (\alpha - 1) \sum_{i=1}^{n} \ln x_i + (\beta - 1) \sum_{i=1}^{n} \ln(1 - x_i).$$

$$(4.6)$$

After that, the optimal values of α and β need to be estimated to maximize $\ln \mathcal{L}(\alpha, \beta \mid x_1, \ldots, x_n)$. Since logarithm is a strictly monotonically increasing function, the maximum value, if it exists, could be computed by

$$\begin{cases} \frac{\partial \ln \mathcal{L}}{\partial \alpha} = 0, \\ \frac{\partial \ln \mathcal{L}}{\partial \beta} = 0. \end{cases} \qquad (4.7)$$

That is

$$g_1(\alpha, \beta) = \psi(\alpha) - \psi(\alpha + \beta) - \frac{1}{n} \sum_{i=1}^{n} \ln x_i = 0, \qquad (4.8)$$

$$g_2(\alpha, \beta) = \psi(\beta) - \psi(\alpha + \beta) - \frac{1}{n} \sum_{i=1}^{n} \ln(1 - x_i) = 0. \qquad (4.9)$$

where $\psi(x)$ is the digamma function defined as

$$\psi(x) = \frac{d}{dx} \ln \Gamma(x) = \frac{\Gamma'(x)}{\Gamma(x)}. \qquad (4.10)$$

There is no closed-form solution to Eqs. (4.8) and (4.9), so Newton–Raphson method is utilized to find the approximate roots. The parameters $\hat{\theta} = [\hat{\alpha} \ \hat{\beta}]$ can be iteratively estimated by Bowman and Shenton [23]

$$\hat{\theta}_{i+1} = \hat{\theta}_i - \frac{\mathbf{g}(\hat{\theta}_i)}{\mathbf{J_g}(\hat{\theta}_i)}, \qquad (4.11)$$

where $\mathbf{g} = [g_1 \ g_2]$, and $\mathbf{J_g}(\hat{\theta}_i)$ is an 2×2 Jacobian matrix defined over the function vector $\mathbf{g}(\hat{\theta}_i)$ defined as

$$\begin{bmatrix} \frac{d\mathbf{g_1}}{d\alpha} & \frac{d\mathbf{g_1}}{d\beta} \\ \frac{d\mathbf{g_2}}{d\alpha} & \frac{d\mathbf{g_2}}{d\beta} \end{bmatrix}, \qquad (4.12)$$

with

$$\frac{d\mathbf{g_1}}{d\alpha} = \psi'(\alpha) - \psi'(\alpha + \beta), \qquad (4.13)$$

$$\frac{d\mathbf{g_1}}{d\beta} = \frac{d\mathbf{g_2}}{d\alpha} = -\psi'(\alpha + \beta), \qquad (4.14)$$

$$\frac{d\mathbf{g_2}}{d\beta} = \psi'(\beta) - \psi'(\alpha + \beta). \qquad (4.15)$$

We consider that such a method has converged if the variation range of $\hat{\theta}$ and $\hat{\beta}$ is less than an acceptable threshold in each successive iteration.

4.3.2.2   ARU Algorithm

By using the exact probability distribution of the anomalous level, the expectation value μ, which is the best indicator of the overall performance of the PMUs during the observation period, can be obtained. Then, the history reputation level of a PMU is defined as

$$T = 1 - \mu = \frac{\beta}{\alpha + \beta}. \qquad (4.16)$$

While, a dependable reputation system should be able to adaptively adjust the reputation values according to dynamic behavioral changes [24]. Therefore, in this chapter, the history reputation level and the subsequent behavior fluctuation of PMUs are considered simultaneously to evaluate their real-time reputation levels. Moreover, as the reputation levels with different behavior observations, adaptive parameters are utilized to allow different impacts. Then, we can define the real-time reputation level of a PMU as

$$T^t = \omega \cdot T_h + (1 - \omega) \cdot T_u^t$$
$$= \omega \cdot \frac{\beta}{\alpha + \beta} + (1 - \omega) \cdot \frac{\lambda_g \cdot N_g^t + 1}{\lambda_g \cdot N_g^t + \lambda_b^t \cdot N_b^t + 1}, \qquad (4.17)$$

where T_h is the history reputation level of a PMU, and T_u^t is the updating reputation level at time instant t. ω is the weight assigned for the history reputation level to assess the importance of history experience to the real-time reputation level, while $1 - \omega$ is for the updating reputation level to assess the impacts of recent performance to the real-time reputation level [25]. N_g^t and N_b^t indicate the cumulative number of observations corresponding to "good" data (not false data) and "bad" data (false data) of a PMU, respectively. Correspondingly, λ_g and λ_b^t are designed as the impact factors for "good" data and "bad" data. From the social perspective, naturally, one needs to take a longer period of time to perform continuous good behaviors for the purpose of a high reputation level establishment, yet only a few bad behaviors would adversely affect the reputation built over time [26]. Thus, the PMUs will be penalized if "bad" data are observed. In our algorithm, λ_b^t is devised relatively larger than λ_g, and λ_b^t will be increased if continuous "bad" data are observed to enlarge the impacts.

Algorithm 4.2 Adaptive reputation updating algorithm

1: **procedure**
2:    **Input:** $N_g^{t-1}, N_b^{t-1}, \lambda_g, \lambda_b^{t-1}, S_b^{t-1}, \tau$
3:    **if** the judgement result of current data is "good" **then**
4:       $N_g^t \leftarrow N_g^{t-1} + 1$;
5:       $S_b^t \leftarrow 0$;
6:    **else**
7:       $N_b^t \leftarrow N_b^{t-1} + 1$;
8:       $S_b^t \leftarrow S_b^{t-1} + 1$;
9:       **if** $S_b^t > 1$ **then**
10:          $\lambda_b^t = \lambda_b^{t-1} \cdot e^\tau$;
11:       **end if**
12:    **end if**
13:    Compute updating reputation level by:
14:    $T_u^t = \dfrac{\lambda_g \cdot N_g^t + 1}{\lambda_g \cdot N_g^t + \lambda_b^t \cdot N_b^t + 1}$,
15:    and the overall reputation level by:
16:    $T^t = \omega \cdot \dfrac{\beta}{\alpha+\beta} + (1 - \omega) \cdot \dfrac{\lambda_g \cdot N_g^t + 1}{\lambda_g \cdot N_g^t + \lambda_b^t \cdot N_b^t + 1}$.
17:    **output:** T^t.
18: **end procedure**

Algorithm 4.2 outlines the ARU procedure, where S_b^t is the number of successive observations of "bad" data. They increase by 1 when corresponding behavior occurs. If continuous "bad" data is observed, the corresponding impact factor λ_b^t will be increased by $\lambda_b^{t-1} \cdot (e^\tau - 1)$, on the contrary, the counter for continues "bad" observations S_b^t will be reset to 0 and the impact factor λ_b^t remains unchanged. Note that τ is initialized as a small value (e.g., 0.0001) in our experiments, and can be adjusted according to different application environments. Here, the time complexity of Algorithm 4.2 is $O(1)$.

With the real-time reputation level of each PMU, it is easy to identify the compromised PMU through testing the following binary hypothesis:

$$\begin{cases} \mathbf{H_0}\text{: PMU } U_j \text{ is compromised,} & \text{if } T_j^t < D_{th} \\ \mathbf{H_1}\text{: PMU } U_j \text{ is not compromised,} & \text{otherwise.} \end{cases} \tag{4.18}$$

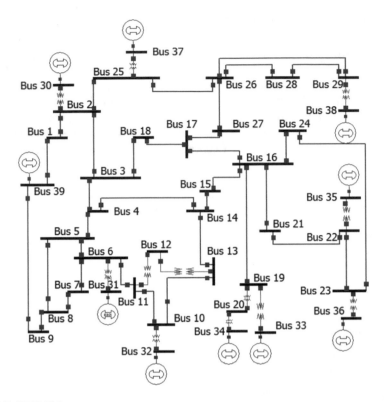

Fig. 4.5 IEEE 39-bus power system

where D_{th} is an acceptable detection threshold. This hypothesis is tested once the reputation level is updated for the purpose of ensuring real-time detection.

4.4 Performance Evaluation

In this section, the effectiveness of our proposed DHCD method, which includes the collaborative FDD process and determination of compromised PMU process, is demonstrated by intensive simulation experiments. The IEEE 39-bus power system shown in Fig. 4.5 is utilized as a benchmark system in our experiments. IEEE 39-bus power system is a well-known New England power system with 10 generators, 39 buses, and 46 transmission lines, which is commonly adopted as a benchmark system to test and verify new schemes [4, 5, 27]. Combined with the PowerWorld simulator [28], real-time, accurate, and precise state information of the power system can be provided. To conduct our experiments, we use the PowerWorld simulator on an IEEE standard 39-bus power system, where we can simulate a lot of scenarios and collect corresponding real-time measurement data from PMUs. We

Table 4.2 Parameter settings

Parameter	Default setting
T_h	0.8
S_b	10
D_{th}	0.6
ω, τ	$\omega = 0.4, \tau = 0.001$
λ_g, λ_b^0	$\lambda_g = 0.1, \lambda_b^0 = 0.5$
Number of PMUs: N	39
Number of samples each test: K	1000
State variables that collected	$\delta, V, L_{Mvar}, L_{MW}$

then use these data to evaluate our proposed DHCD method in MATLAB. Table 4.2 lists the key parameters.

4.4.1 Efficacy of FDD Algorithm

Two groups of simulation experiments are carried out in this section. It can be drawn from the first group that only one piece of the four rule specifications is violated (with a single "1" in R_j^t). On the contrary, as shown in the second group, multiple pieces of the four rule specifications are violated (with multiple "1"s in R_j^t). In addition, as shown in Fig. 4.6, we divide each group into four different cases: (a) single, (b) sparse, (c) random, and (d) dense, representing four distribution types of false measurement data. Specifically, case (a) describes that only single PMU is inserted with false measurement data; case (b) describes that multiple sparsely distributed PMUs are inserted with false measurement data; case (c) describes that multiple randomly distributed PMUs are injected with false measurement data; and case (d) describes that multiple densely distributed PMUs are injected with false measurement data.

As shown in Tables 4.3 and 4.4, the simulation results are based on the detection rate and average number of iterations of the FDD algorithm for detecting single violation of rules and multiple violations of false measurement data respectively. It can be observed that the proposed FDD algorithm can easily achieve 100% detection rate in the detection of singly and sparsely distributed PMUs compromised by attackers. The proposed FDD can achieve high detection rate but not 100% in terms of detecting randomly or densely distributed PMUs compromised by adversaries. This is because, collaborating with FDD can detect false data more accurately in most cases if these PMUs which corresponded are located near the inner regions of the grid. Starting from the peripheral PMUs at the first iteration to the inner PMUs at the subsequent iterations, the anomalies can be detected. However, only peripheral PMUs near the inner regions can be identified if these anomalous PMUs are concentrated in the edge area of the grid in some extreme and rare cases. The peripheral anomalous PMUs can be identified and their connections to other PMUs

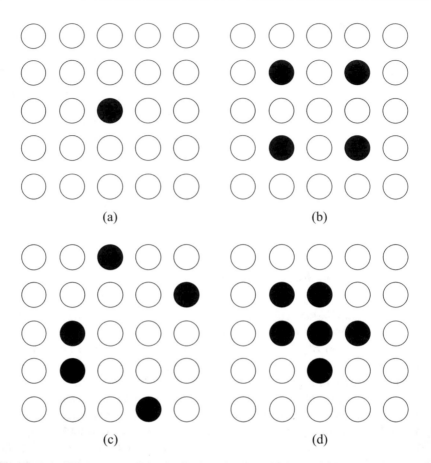

Fig. 4.6 Four different cases of the distribution of PMUs with inserted false measurement data: single, sparse, random, and dense. (**a**) Single. (**b**) sparse. (**c**) Random. (**d**) Dense

Table 4.3 The detection rate and the average iterations of FDD algorithm with single rule violated false measurement data under four different distribution types. The number of PMUs with false measurement data is 6

Distribution type	Detection rate	Average iterations
Single	100.0%	1.000
Sparse	100.0%	1.000
Random	97.10%	1.173
Dense	80.40%	2.071

removed after the first or two iterations. Thus, other anomalous PMUs for edge regions can be isolated, with only the anomalous neighboring PMUs. By showing the same results R_i^t, they can collude with each other to mutually protect each other. As such extreme cases are more likely occur in dense distribution type simulation

Table 4.4 The detection rate and the average iterations of FDD algorithm with multiple rules violated false measurement data under four different distribution types. The number of PMUs with false measurement data is 6

Distribution type	Detection rate	Average iterations
Single	100.0%	1.000
Sparse	100.0%	1.000
Random	97.90%	1.107
Dense	93.70%	1.520

experiments, the dense type in both group one and group two has relatively lower detection rate.

In group one and group two, the average iterations for either singly or sparsely distributed PMU(s) compromised by attackers are 1.000 due to these two types of injected exception data can be easily identified by cooperative detection with only one iteration. The average iterations are 1.173 and 1.107 for the two groups, respectively, in random distribution type. It means that in some cases, instead of one round, the extra one to two rounds FDD are required to detect the false measurement data. In our simulation experiments, we set the maximum iteration of executing FDD as 5 for undetected false data. As for the densely distribution type, the average iterations are 2.071 and 1.0520, respectively. That is, in order to detect the inner false data, it requires extra FDD iterations in more cases compared with random distribution type.

Interestingly, compared with group one simulations, a higher or equal detection rate can be achieved with fewer average iterations in group two simulations can be concluded from numerical simulation results. The reason is that the false data can be detected by our FDD algorithm when at least one rule is violated, so in the second group, FDD is more likely to detect abnormal data.

Furthermore, we studied the relationship between the average iterations and the number of compromised PMUs under random distribution type as shown in Fig. 4.7, and the corresponding detection rate as well in Fig. 4.8. Clearly, as the increase in the number of compromised PMUs, the value of the average iterations increases and eventually up to 5, i.e., the preset maximum value. Correspondingly, the value of the detection rate drops from 1 to 0 while the number of PMUs with false data increases. Similar results can be observed, in this case, the value of the average number of iterations and the detection rate of multiple rule violations are better than those of a single rule violation.

4.4.2   Identification of Compromised PMUs with Our Reputation System

Several critical parameters can affect our reputation system's performance, which are shown below: (1) ω, the weight assigned for the history reputation level; (2)

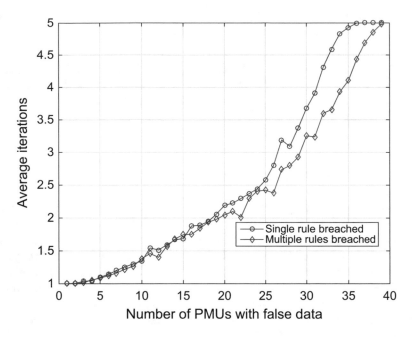

Fig. 4.7 The average iterations needed for FDD algorithm vs. different numbers of PMUs with false measurement data

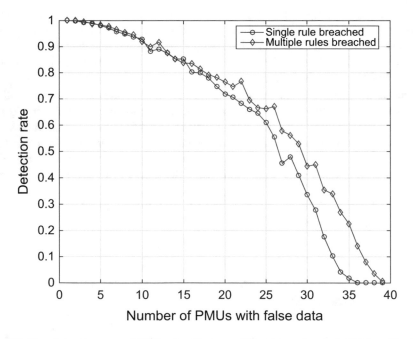

Fig. 4.8 The detection rate of FDD algorithm vs. different numbers of PMUs with false measurement data

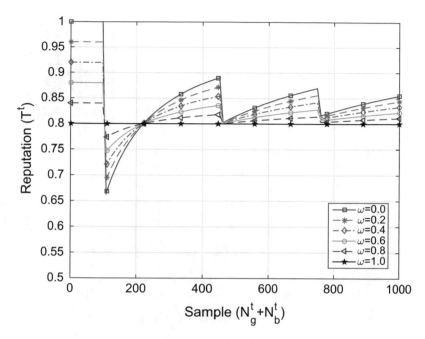

Fig. 4.9 The reputation level of a PMU under various values of ω ($T_h = 0.8$, $D_{th} = 0.6$, $S_b = 10$, $\lambda_b^0 = 0.5$)

D_{th}, the detection threshold; (3) λ_b, the impact factor; and (4) S_b^t, the number of successive observations of "bad" data.

It can be obtained from Fig. 4.9 that it shows the fluctuations of a PMU's reputation level under various values of ω. Three FmDI events, each lasting 10 samples, are inserted into the PMU's measurement data. It can be observed from this figure that, the higher the ω is, the more the current reputation level T^t relies on its history value T_h. Particularly, $\omega = 0.0$ indicates that $T^t = T_h$, and $\omega = 1.0$ denotes that $T^t = T_u^t$.

Figure 4.10 shows the fluctuations of a PMU's reputation level under various values of D_{th}. Six FmDI events, each lasting 10 samples, are inserted into the PMU's measurement data. It can be observed from this figure that a higher D_{th}s hold a lower tolerance to PMUs' "bad" behaviors, while lower D_{th}s have higher tolerance to PMUs' "bad" behaviors. In other words, higher D_{th}s are more sensitive than lower D_{th}s. For instance, when $D_{th} = 0.65$, our reputation system raises an alarm when the first FmDI event is inserted.

The relationship between the reputation level and the λ_b^0 can be observed in Fig. 4.11. Three FmDI events, each lasting 10 samples, are inserted into the PMU's measurement data. It can be seen that, the more adverse the consequence of penalty to the reputation level with higher λ_b^0, which means that the reputation level decreases noticeably. It can be seen that, as the λ_b^0 increases, the reputation level

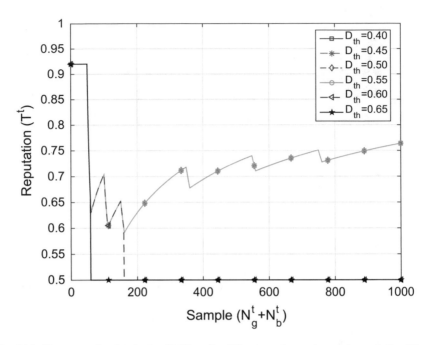

Fig. 4.10 The reputation level of a PMU under different under various values of D_{th} ($T_h = 0.8$, $\omega = 0.4$, $S_b = 10$, $\lambda_b^0 = 0.5$)

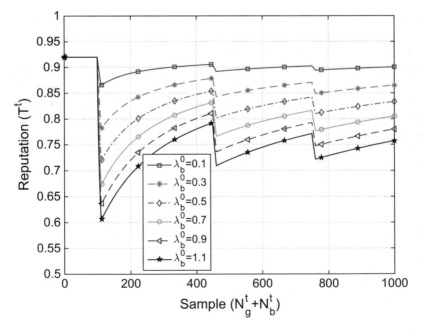

Fig. 4.11 The reputation level of a PMU under various values of λ_b^0 ($T_h = 0.8$, $\omega = 0.4$, $D_{th} = 0.6$, $S_b = 10$)

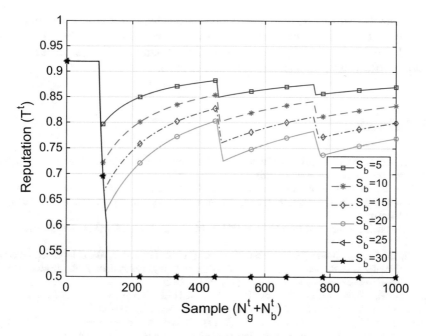

Fig. 4.12 The reputation level of a PMU under various values of S_b ($T_h = 0.8, \omega = 0.4, D_{th} = 0.6, \lambda_b^0 = 0.5$)

drops more significantly. This is because the higher the λ_b^0, the consequences of penalty are less favorable to reputation.

As shown in Fig. 4.12, we plot a similar relationship between the reputation level and the S_b. Besides, we insert three FmDI events but different lengths into the PMU's measurement data. What is similar to Fig. 4.11 is that this figure shows the more significance the penalty has on the reputation level with larger S_b, because large S_b brings about more times of λ_b^t adjustment, i.e., $\lambda_b^t = \lambda_b^{t-1} * e^\tau$. For example, with $D_{th} = 0.6$, the reputation level drops dramatically below D_{th} if $S_b = 30$.

4.5 Summary

In this chapter, a novel DHCD method is proposed to detect and relatively reduce FmDI attacks in smart grid CPS. To be specific, we design a rule specification based real-time collaborative detection system to detect false measurement data. Moreover, in order to detect PMUs with false measurement data, we present a new reputation system with an ARU algorithm that can assess the overall running status of the PMUs. After that, the effectiveness of the proposed method has been demonstrated using simulations of the IEEE 39-bus power system.

As mentioned earlier, the primary goal of our approach is to detect malicious activity that can cause anomalies in the measurement data. In our future work, we would like to extend our proposed method to capture power system faults (e.g., voltage disturbance, open circuit, and short circuit).

References

1. Falliere, N., Murchu, L. O., & Chien, E. (2011). W32. Stuxnet dossier. *White paper, Symantec Corporate, Security Response, 5*(6), 29.
2. Shin, S., & Gu, G. (2010). Conficker and beyond: A large-scale empirical study. In *Proceedings of the 26th Annual Computer Security Applications Conference (ACSAC), Austin, 2010* (pp. 151–160).
3. Gorman, S., Dreazen, Y. J., & Cole, A. (2009). Insurgents hack US drones. *Wall Street Journal, 17*.
4. Li, B., Lu, R., Wang, W., & Choo, K.-K. R. (2016). DDOA: A Dirichlet-based detection scheme for opportunistic attacks in smart grid cyber-physical system. *IEEE Transactions on Information Forensics and Security, 11*(11), 2415–2425.
5. Bao, H., Lu, R., Li, B., & Deng, R. (2016). BLITHE: Behavior rule based insider threat detection for smart grid. *EEE Internet of Things Journal, 3*(2), 190–205.
6. Fang, H., Xu, L., & Choo, K.-K. R. (2017). Stackelberg game based relay selection for physical layer security and energy efficiency enhancement in cognitive radio networks. *Applied Mathematics and Computation, 296*, 153–167.
7. Chen, J., Shi, L., Cheng, P., & Zhang, H. (2015). Optimal denial-of-service attack scheduling with energy constraint. *IEEE Transactions on Automatic Control, 60*(11), 3023–3028.
8. Handschin, E., Schweppe, F., Kohlas, J., & Fiechter, A. (1975). Bad data analysis for power system state estimation. *IEEE Transactions on Power Apparatus and Systems, 94*(2), 329–337.
9. Merrill, H. M., & Schweppe, F. C. (1971). Bad data suppression in power system static state estimation. *IEEE Transactions on Power Apparatus and Systems,6*, 2718–2725.
10. Chen, J., & Abur, A. (2006). Placement of PMUs to enable bad data detection in state estimation. *IEEE Transactions on Power Apparatus and Systems, 21*(4), 1608–1615.
11. Kotiuga, W. W., & Vidyasagar, M. (1982). Bad data rejection properties of weighted least absolute value techniques applied to static state estimation. *IEEE Transactions on Power Apparatus and Systems, 4*, 844–853.
12. Cutsem, T. V., Ribbens-Pavell, M., & Mili, L. (1984). Hypothesis testing identification: A new method for bad data analysis in power system state estimation," *IEEE Transactions on Power Apparatus and Systems, 11*, 3239–3252.
13. Liu, Y., Ning, P., & Reiter, M. K. (2011). False data injection attacks against state estimation in electric power grids. *ACM Transactions on Information and System Security, 14*(1), 13.
14. Baran, M. E., & Kelley, A. W. (1994). State estimation for real-time monitoring of distribution systems. *IEEE Transactions on Power Apparatus and Systems, 9*(3), 1601–1609.
15. Nordman, M. M., & Lehtonen, M. (2005). Distributed agent-based state estimation for electrical distribution networks. *IEEE Transactions on Power Apparatus and Systems, 20*(2), 652–658.
16. Li, W. (2014). *Risk assessment of power systems: Models, methods, and applications*. London: Wiley.
17. Qiu, M., Gao, W., Chen, M., Niu, J.-W., & Zhang, L. (2011). Energy efficient security algorithm for power grid wide area monitoring system. *IEEE Transactions on Smart Grid, 2*(4), 715–723.
18. Qiu, M., Su, H., Chen, M., Ming, Z., & Yang, L. T. (2012). Balance of security strength and energy for a PMU monitoring system in smart grid. *IEEE Communications Magazine, 50*(5), 142–149.

19. Castiglione, A., Pizzolante, R., Esposito, C., De Santis, A., Palmieri, F., & Castiglione, A. (2017). A collaborative clinical analysis service based on theory of evidence, fuzzy linguistic sets and prospect theory and its application to craniofacial disorders in infants. *Future Generation Computing Systems, 67*, 230–241.
20. Patel, A., Alhussian, H., Pedersen, J. M., Bounabat, B., Júnior, J. C., & Katsikas, S. (2017). A nifty collaborative intrusion detection and prevention architecture for smart grid ecosystems. *Computers & Security, 64*, 92–109.
21. Esposito, C., Castiglione, A., Palmieri, F., & Ficco, M. (2017). Trust management for distributed heterogeneous systems by using linguistic term sets and hierarchies, aggregation operators and mechanism design. *Future Generation Computer Systems, 74*, 325–336.
22. Premarathne, U. S., Khalil, I., & Atiquzzaman, M. (2016). Trust based reliable transmissions strategies for smart home energy consumption management in cognitive radio based smart grid. *Ad Hoc Network, 41*, 15–29.
23. Bowman, K., & Shenton, L. (1992). Parameter estimation for the Beta distribution. *Journal of Statistical Computation and Simulation, 43*(3–4), 217–228.
24. Srivatsa, M., Xiong, L., & Liu, L. (2005). TrustGuard: Countering vulnerabilities in reputation management for decentralized overlay networks. In *Proceedings of the 14th International Conference on World Wide Web (WWW), Chiba, 2005* (pp. 422–431).
25. Mármol, F. G., & Pérez, G. M. (2012). TRIP, a trust and reputation infrastructure-based proposal for vehicular ad hoc networks. *Journal of Network and Computer Applications, 35*(3), 934–941.
26. Sun, Y. L., Han, Z., Yu, W., & Liu, K. R. (2006). A trust evaluation framework in distributed networks: Vulnerability analysis and defense against attacks. In *Proceedings of the 25th IEEE INFOCOM, Barcelona, 2006* (pp. 1–13).
27. Zhang, D., Li, S., Zeng, P., & Zang, C. (2014). Optimal microgrid control and power-flow study with different bidding policies by using powerworld simulator. *IEEE Transactions on Sustainable Energy, 5*(1), 282–292.
28. PowerWorld. (2018). Retrieved June 28, 2018 from https://www.powerworld.com

Chapter 5
DDOA: A Dirichlet-Based Detection Scheme for Opportunistic Attacks

In future smart grid cyber-physical systems, a hierarchical control paradigm will be widely accepted. In this paradigm, decentralized local agents (LAs) may potentially be compromised by opportunistic attackers with a purpose of manipulating the real-time electricity prices and then gaining illicit financial profits. In this chapter, we propose a Dirichlet-based detection scheme (DDOA) to contain such FcDI cyberattacks. Specifically, we build a Dirichlet-based probabilistic model to evaluate the reputations of LAs. The initial reputation levels of these LAs are trained based on the historical operating observations. Then, an adaptive cyber threat detection algorithm along with a reputation incentive mechanism is designed to identify opportunistic attackers. Extensive experiments on the IEEE 39-bus power system with the PowerWorld simulator demonstrate the utility of our DDOA scheme.

5.1 Introduction

The rapid advancement of ICTs makes smart grid cyber-physical system a commonplace. Smart grid cyber-physical system is a geographically large-scale and complex interconnected power infrastructure that spans across multiple regions. To ensure the high reliability and robustness of the critical physical infrastructure, real-time monitoring of the whole power grid, accurate data analytics, and automated control are highly critical. Usually, data analytics is conducted by a state estimator at the system control center [1–3]. Nevertheless, with the increasing interconnections, dynamics, and nonlinearity, real-time data analytics inevitably create heavy computational burdens and complexities on the control center [4]. If this cannot be properly managed, the control center's operating efficiency will be negatively affected, leading to cascading impacts—e.g., affecting the reliability as well as the robustness of the power system and eventually destroying it. One effective solution to meet the strict computational requirements on the control center presented in the

© Springer Nature Switzerland AG 2020   99
B. Li et al., *Detection of False Data Injection Attacks in Smart Grid
Cyber-Physical Systems*, Wireless Networks,
https://doi.org/10.1007/978-3-030-58672-0_5

literature is the hierarchical control paradigm. In this framework, dispersived LAs conduct real-time data analytics in their respective local regions [5, 6].

Although a hierarchical paradigm is able to availably descend the computational costs of the control center, it may lead to unintended security consequences [4]. For instance, in the existing centralized power grid, it is easier secure the control center; thus, control center is generally considered as a fully trusted party. However, in a hierarchical framework, it is almost impossible to anticipate that all distributed LAs can be protected to an identical level as the control center.

The wide integration of power systems with ICTs has also led to an expanded attack landscape. For instance, the inherent vulnerabilities in existing power grids or devices and/or entry points that connected to the power grid may be employed by cyber attackers. As per the newsletter of the Industrial Control Systems Cyber Emergency Response Team (ICS-CERT), in Fiscal Year 2015 (i.e., 1 October 2014 to 30 September 2015), ICS-CERT of the U.S. Department of Homeland Security has reported a total number of 295 cybersecurity incidents relevant to critical infrastructures. It is worth noting that the energy sector is the second largest sector with the most targeted national critical infrastructure [7]. The increasing importance of cyber threats to cyber-physical systems are further evidenced by recent cyberattacks (e.g., on a German steel mill which destroyed a blast furnace [8]) and cyber attempts (e.g., ISIS is reported with an attempt to attack US electric power utilities in purpose of stealing confidential power grid information and mount terrorist attacks [9]). A successful cyberattack may potentially overwhelm and paralyze the country's widely and deeply interconnected national critical infrastructures and, therefore, lead to severe social panics and unrests.

Particularly, cybersecurity issues of smart electricity markets have recently attracted wide attention of researchers from both security community as well as the power community [10–12]. Unfortunately, we notice that existing efforts are devoted with a concentration on mitigating false data injection attacks (i.e., cyber attackers falsify measurement data to "blind" the system operator in purpose of manipulating real-time electricity prices [13]). Generally speaking, it is assumed that attackers have the knowledge of system topology and configurations, and are capable of simultaneously falsifying a set of PMU reported measurement data at will. Kosut et al. in 2011 studied multiple attacker's strategies and their relevant countermeasures for data integrity attackers in smart grids [14]. Xie and Esmalifalak et al. also investigated FmDI attacks in electricity markets, which are able to manipulate locational electricity prices [11, 15, 16]. As we can see, these studies all concentrated on a centralized power grid model. With the increasing demands on interconnectivity between systems in future smart grids, recent research focus have shifted to security in hierarchical smart grids (see [4, 17–19]). For instance, Li [19] proposed a decentralized quick detection method for FmDI attacks in smart grids. Vukovic [4] carefully analyzed the cybersecurity issues in distributed power grids and designed a method to detect and mitigate data integrity attacks in the power grids.

Following this, we notice that most of the existing efforts were guided to those data integrity attacks that may cause power system outages. In addition to politically-, criminally-, and ideologically motivated attacks, cyber attackers may be interested in manipulating smart electricity markets for illicit financial profits [13, 15, 20]. Opportunistic attackers [1, 21] are one such example, which usually are initiated by inside attackers. Concretely, rather than compromising a set of PMUs to forge the measurement data, opportunistic attackers aim to manipulate the electricity prices by compromising some key intelligent control device which is able to determine the real-time electricity prices, say the LA. Compromised LAs are capable of issuing fake commands to its local generators, transformers, and distributors to change the normal demand-supply balances, which will further affect the electricity prices at each bus. If these attackers colluded with some key participants in electricity markets, such as the power suppliers or utilities, they may obtain a large amount of illicit financial gains during the rapid fluctuations of electricity prices [15].

In most cases, opportunistic attackers prefer not to cause any physical damage to power grids, it has become a challenging task now for traditional IDS to detect. In addition, opportunistic attackers can flexibly adapt their strategies, for example, the probability to mount an attack whenever there is an opportunity, based on the current system noise level to avoid being detected [1]. Thus, in order to detect the abnormality of any potential compromised LAs, an effective solution is to observe and assess their operating behaviors (i.e., commands issued and corresponding variable status) over a long period of time. In this chapter, in order to mitigate such opportunistic attacks, we present a novel Dirichlet-based detection scheme (called DDOA). This scheme permits the system control center to effectively detect compromised LAs in the way of observing and assessing their operating behaviors. The main contributions of this chapter are in threefold:

- First, we propose a three-tier hierarchical framework for the smart grid infrastruc-tures, which is devised to effectively reduce the control center's computational burdens. By employing this framework, guaranteeing a high reliable and robust future smart grids become possible.
- Second, we are the pioneers, to our best knowledge, to study such opportunistic attacks on smart electricity markets, and establish a Dirichlet-based reputation assessment model to assess the performance of each LA through consistently observing their operating behaviors.
- Last, we design an adaptive detection scheme with a reputation incentive mechanism for future smart grids, which is capable of effectively and accurately detecting possible opportunistic attackers who aim to manipulate the electricity prices from the smart electricity markets. In addition, two levels of detection thresholds are used in the proposed DDOA scheme, which is beneficial for differentiating malicious activities from usual system faults in future smart grids.

The remainder of this chapter is organized as follows. Section 5.2 presents the system model, the threat model, and our design goals. In Sect. 5.3, we introduce the preliminaries needed in the understanding of this chapter. We elaborate our proposed Dirichlet-based reputation model and detection scheme in Sect. 5.4, and present the performance evaluation in Sect. 5.5. Section 5.6 concludes the chapter.

5.2   Models and Design Goals

In this section, we introduce both the system and threat models, as well as our design goals.

5.2.1   System Model

In our system model, we consider a hierarchical flocking-based framework for future smart grids (see Fig. 5.1). This framework is composed of three tiers, that is, the PMU tier at the lowermost, the LA tier at the intermediate, and the control center tier at the uppermost. The roles and responsibilities of each entity in these tiers are illustrated below:

- PMUs, located at power buses and generators over the whole power grid, are geographically flocked, which forms into several flockings. They are responsible for collecting real-time measurement data of system operating states in each flocking area, such as the power loads **L**, power generations **G**, and line power flows **F**. Then, these collected measurement data are reported to the PDC located in the LA tier.
- The LA in a flocking area is composed of a PDC, a state estimator, as well as a local controller, which is responsible for analyzing the real-time system states of this local area, and delivering the reported data to the uppermost tier

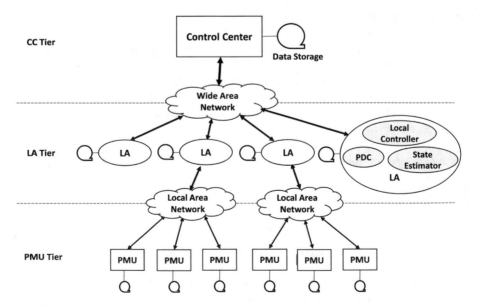

Fig. 5.1 Three-tier hierarchical flocking-based framework for future smart grids

of the control center if in need. Concretely, the PDC aggregates the measurement data reported by PMUs; the state estimator is able to estimate the real system states in this flocking area; and then the local controller is in charge of analyzing the estimated system states, determining the locational marginal prices (LMPs), and issuing feedback control commands to the local generators, distributors, and transformers, etc.

• The control center takes the responsibility of storing and analyzing the measurement data for multiple applications, such as the state estimation, contingencies analysis, as well as event diagnostics, etc. Furthermore, in our scheme, the control center also monitors and assesses the reputation levels of the underlying LAs in purpose of identifying any abnormal behaviors.

Note that, in this chapter, it is assumed that both the control center and LAs employ the state estimation to, respectively, estimate the system states of either the entire power grid or a specific local region. Particularly, the control center carries out state estimation in a relatively low frequency, in order to reduce the computational burden (see Sect. 5.4.3).

5.2.2 Threat Model

Unlike traditional centralized power grids, as aforementioned, the future smart grids will transfer the real-time monitoring, data analysis, and real-time control tasks from the control center to its underlying LAs. It is intuitive to assume that the control center is a fully trusted entity, while the LAs are not who might be easily compromised by strong attackers. In this work, we assume that PMUs are honest parties, which is to say, the measurement data reported by PMUs are always without malicious falsifications.

If the attackers can successfully compromise an LA, they are able to mount FcDI attacks by issuing forged control commands to the regional power generators, transformers, and distributors, in purpose of manipulating normal demand-response balances. These acts may lead to alterations of the LMPs in a specific or some targeted areas. Since it is a premeditated activity, the FcDI attackers can obtain a large amount of financial gains by exploiting such price changes. For instance, the attackers may collude with some other parties in the smart electricity bidding markets and buy a considerably amount of electricity energy at a very low price either by an FcDI attack or not; once the price has been normally or artificially raised up, they may choose to sell these pre-purchased electricity to users or other parties in the markets.

As Fig. 5.2 shows, an example of the electricity prices distribution under normal operating conditions on the standard IEEE 39-bus power system is described by a contouring map. Note that, multiple colors are used to denote different demand-supply balances in each area. In cases where malicious FcDI attacks happen, these normal balances and correspondingly, the electricity prices can be changed

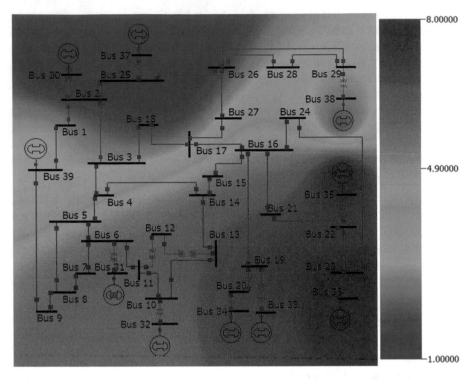

Fig. 5.2 The contouring map of electricity price distribution on IEEE 39-bus power system

intentionally by these attackers. We broadly categorize these cyberattacks into random, reckless, and opportunistic attacks.

1. Random attacks: this type of cyberattacks are mounted in a random mode but with a determined attack probability $P_a \in [0, 1]$. These cyberattacks can be easier identified by employing conventional IDSs or intrusion prevention systems (IPSs), in that they have a fixed attacks probability.
2. Reckless attacks: this type of cyberattacks are mounted in an ad-hoc mode, that is to say, once there is an opportunity, these attackers attempt to initiate an attack without any hesitation or careful planning. In consequence, these reckless attackers are always the easiest to be detected.
3. Opportunistic attacks: this type of cyberattacks are mounted on the basis of the current system noise, the attack probability of which is given by $P_a = C \cdot P_n^{\varepsilon}$, where C denotes a constant number, and ε is a scalar for the current system noise P_n. It is worth noting that, $\varepsilon > 1$ implies the cases where conservative opportunistic attackers initiate such attacks, while $\varepsilon < 1$ implies the cases where aggressive attackers initiate such attacks. As we see, the larger the current system noise is, the higher the attack probability is.

Note that, opportunistic attackers should be the trickiest attackers among the above-mentioned, in that they are able to adapt their attack probabilities with respect to the current system noise. In this way, it becomes considerably challenging to detect such FcDI attackers by employing the conventional detection schemes. In this chapter, our goal is to design an effective scheme to identify as well as detect FcDI attackers, particularly for such opportunistic attackers.

5.2.3 Design Goals

The main goal of our proposed scheme is to design an available and accurate measurement to identify and detect opportunistic attacks in future smart grids, which comprises the following three sub-goals:

1. Future power grids are envisioned to be hierarchical systems, owing to their capabilities of ensuring high stability, efficiency, and reliability of power grids in cases with ever-increasing electricity demands, fusion of renewable energy resources, as well as multiple applications. In this way, we design a three-tier hierarchical control framework for future power grids in purpose of supporting these critical requirements.
2. LAs play a significant role in distributed power grid areas. It is, therefore, of great importance to guarantee their functionalities. As above-mentioned, unlike the control center, LAs are not fully trusted parties. Thus, we need to efficiently and accurately monitor and assess their behaviors to further ensure their security. In this chapter, we propose a Dirichlet-based reputation model to accurately assess LA's operating behaviors.
3. In order to consistently monitor all LAs' operating behaviors and identify any potential abnormal behaviors, we design an effective detection scheme based on the proposed Dirichlet-based reputation model to detect potential LAs that are possibly compromised by opportunistic attackers. Furthermore, the effectiveness of the proposed DDOA scheme is demonstrated by using the real-time measurement data collected in a PowerWorld simulator.

5.3 Preliminaries

In this part, we brief some important preliminaries needed in the help for the understanding the rest of the parts of this chapter.

5.3.1   State Estimation

Although the basic concepts of state estimation have been introduced in Sect. 2.1, we provide additional details of what kinds of data are measured and estimated. As shown in Table 1.1, PMUs can measure the values of a myriad of state variables. The measurement data can be classified into three main types, including power generations $\mathbf{z}_G$, power loads $\mathbf{z}_L$, line power flows $\mathbf{z}_F$.

As per the DC power flow model, the estimated system states are given by

$$\hat{\mathbf{x}} = \begin{bmatrix} \hat{\mathbf{x}}_V \\ \hat{\mathbf{x}}_\theta \end{bmatrix} = \mathbf{H} \mathbf{\Lambda} \begin{bmatrix} \mathbf{z}_G \\ \mathbf{z}_L \\ \mathbf{z}_F \end{bmatrix}. \tag{5.1}$$

Usually, the estimated system states only comprise voltage magnitudes $\hat{\mathbf{x}}_V$ and phase angles $\hat{\mathbf{x}}_\theta$. While it is easy to further calculate the estimated power generations $\hat{\mathbf{x}}_G$, power loads $\hat{\mathbf{x}}_L$, line power flows $\hat{\mathbf{x}}_F$ by using $\mathbf{x} = [\hat{\mathbf{x}}_V, \hat{\mathbf{x}}_\theta]$ [13], i.e.,

$$\begin{bmatrix} \hat{\mathbf{x}}_G \\ \hat{\mathbf{x}}_L \\ \hat{\mathbf{x}}_F \end{bmatrix} = F \begin{bmatrix} \hat{\mathbf{x}}_V \\ \hat{\mathbf{x}}_\theta \end{bmatrix} = F \mathbf{H} \mathbf{\Lambda} \begin{bmatrix} \mathbf{z}_G \\ \mathbf{z}_L \\ \mathbf{z}_F \end{bmatrix}, \tag{5.2}$$

where F is a matrix relating the voltage magnitudes and phase angles to power generations, power loads, and line power flows.

5.3.2   Real-Time LMPs

In future smart electricity markets, the real-time LMP in an LA area will be computed on the basis of the estimated locational real-time system states. The LMP is expressed as the per cost to serve the next unit increment of power loads, e.g., 1 MWh, at each bus, by jointly considering real power loads, power generations, and line power flows subject to power limits of transmission lines [22].

The least square linear optimization problem is reduced to incremental linear optimization problem by using the estimated system state (see Eq. (5.3)). The objective is to minimize the cost function subject to the constraints of power balances, power generation bounds, power transaction bounds, and any other power transmission constraints that exist on the current power systems. This optimization problem is given below:

$$\min \quad \mathcal{J} = \sum C_i(\Delta G_i) - \sum C_j(\Delta L_j)$$

$$s.t. \quad \sum \Delta G_i - \sum \Delta L_j = 0$$

$$\Delta G_i^{min} \leq \Delta G_i \leq \Delta G_i^{max} \tag{5.3}$$

$$\Delta L_i^{min} \leq \Delta L_i \leq \Delta L_i^{max}$$

$$A_{ik}\Delta G_i + D_{jk}\Delta L_j \leq 0,$$

where C_i and C_j, respectively, denote the calculated real-time offer of generator i and real-time bid of load j, [22]. A_{ik} is a matrix encompassing the shift factors for generation bus i on the binding transmission constraints (k), while D_{jk} denotes a matrix encompassing the shift factors for load bus j on the binding transmission constraints (k). In this way, the LMP values at each bus can be calculated, which is given by

$$LMP_i = \lambda - \sum A_{ik} * SP_k, \tag{5.4}$$

where λ denotes the locational marginal price of generation at the reference bus [23], A_{ik} is a shift factor for bus i on binding constraint k, and SP_k denotes the shadow price of constraint k.

5.3.3 Dirichlet Distribution

The Dirichlet distribution [24] belongs to the family of continuous multivariate probability distributions, with parameters of a vector $\boldsymbol{\alpha}$ of positive real numbers. Let $X = \{x_1, x_2, \ldots, x_k\}$ be a discrete random variable, where $x_i > 0$ for $i = 1, 2, \ldots, k$ and $\sum_{i=1}^{k} x_i = 1$. Suppose that $\boldsymbol{\alpha} = [\alpha_1, \alpha_2, \ldots, \alpha_k]$ with $\alpha_i > 0$ for all i from 1 to k, and let $\alpha_0 = \sum_{i=1}^{k} \alpha_i$. Then, X can be said a Dirichlet distribution with parameters of $\boldsymbol{\alpha}$, which is expressed as $X \sim Dir(\boldsymbol{\alpha})$. The probability density function (PDF) can then be given as

$$f(X; \boldsymbol{\alpha}) = \frac{1}{B(\boldsymbol{\alpha})} \prod_{i=1}^{k} x_i^{\alpha_i - 1} = \frac{\Gamma(\alpha_0)}{\prod_{i=1}^{k} \Gamma(\alpha_i)} \prod_{i=1}^{k} x_i^{\alpha_i - 1}, \tag{5.5}$$

where $B(\cdot)$ denotes a Beta function and $\Gamma(\cdot)$ denotes a Gamma function. Its expectation and variance of $X = x_i$ are, respectively, expressed as

$$E[x_i] = \frac{\alpha_i}{\alpha_0}, \quad \text{Var}[x_i] = \frac{\alpha_i(\alpha_0 - \alpha_i)}{\alpha_0^2(\alpha_0 + 1)}. \tag{5.6}$$

5.4   The Proposed DDOA Scheme

In this part, we introduce the details of the proposed DDOA scheme comprising three partitions: behavior rule specifications, Dirichlet-based reputation model, and detailed description of the proposed DDOA scheme.

5.4.1   Behavior Rule Specifications

Smart grids are highly interconnected cyber-physical systems. The physical behaviors, such as the operations and variable states, of intelligent electronic devices fully reflect the control commands from the cyber space. In this case, it is natural and reliable to identify the abnormalities in the cyber space by evaluating the behaviors of physical devices. On the other hand, complex interconnections and balances in power grids can lead to various inter-constraints between the state variables, which can be employed to define a line of rule specifications for the power grid managers, such as the control units. In this chapter, following this, we define a set of behavior rule specifications for LAs that they must obey in the normal operating conditions (see Table 5.1 for some examples). This is helpful for identifying any operating abnormality in power grids.

For example, as to the first rule $R1$, G_i^t is the measured power generation of generator i at time t, while $\hat{G}_i^t$ is the corresponding expected power generation value. $R1$ depicts the difference between the absolute measured power generation value and the absolute expected power generation value should be constrained to a specified safe threshold τ_G. In the proposed scheme, due to the fact that the control center is the fully trusted party, we define the expected values by the values estimated, using state estimation, by the control center. In real-world applications, besides rule $R1$, the value of G_i^t itself should also be limited to a safe range,

Table 5.1 Rule specifications

Index	Rule	Description
$R1$	$\lvert G_i^t - \hat{G}_i^t \rvert \leq \tau_G$	The absolute difference of G between the measured and expected values should be less than a safe threshold τ_G
$R2$	$\lvert L_i^t - \hat{L}_i^t \rvert \leq \tau_L$	The absolute difference of L between the measured and expected values should be less than a safe threshold τ_L
$R3$	$\lvert F_i^t - \hat{F}_i^t \rvert \leq \tau_F$	The absolute difference of F between the measured and expected values should be less than a safe threshold τ_F
$R4$	$\tau_P^{min} \leq G_i^t \leq \tau_G^{max}$	The value of G itself should be constrained within a specified safe range $[\tau_G^{min}, \tau_G^{max}]$
$R5$	$\tau_L^{min} \leq L_i^t \leq \tau_L^{max}$	The value of L itself should be constrained within a specified safe range $[\tau_L^{min}, \tau_L^{max}]$
$R6$	$\tau_F^{min} \leq F_i^t \leq \tau_F^{max}$	The value of F itself should be constrained within a specified safe range $[\tau_F^{min}, \tau_F^{max}]$

e.g., $[\tau_G^{min}, \tau_G^{max}]$, as described by $R4$. Likewise, similar rules can also be defined, respectively, for power loads **L** and power line flows **F** (see other rules in Table 5.1).

As mentioned above, the measurements of the state variables are reflecting the LA's behaviors. It is, therefore, logical to claim that deviations of these rule specifications can be used to identify abnormalities. Nevertheless, a single deviation of rule specifications may not sufficiently imply that an LA is compromised, as system noises may also lead to deviations. As a result, in this chapter, we employ a conjunctive form of rule specifications and consistent long-term observation of the conjunctive rule specifications, in order to achieve effective and accurate assessment of LA's behaviors. The conjunctive rule specification $\mathcal{R}$, which combines all of these specified rules, can be given by

$$\mathcal{R} = R1 \cup R2 \cup R3 \cup R4 \cup R5 \cup R6. \tag{5.7}$$

Note that "1" is used to simply denote the non-compliance of a rule specification, while "0" is used to denote the compliance. In this way, the compliance condition of $\mathcal{R}$ can be denoted by a binary sequence. For instance, "100010" represents that rules $R1$ and $R5$ are non-compliant, while the rest rules are compliant. In particular, the full compliance of all the six conjunctive rules is denoted by "000000," which is defined as our reference sequence seq_{ref} for rule compliance.

To facilitate further discussions, we define the *compliance level* of each measurement data as follows:

$$\rho = 1 - dist(seq, seq_{ref}), \tag{5.8}$$

where $dist(\cdot, \cdot)$ is a function denoting the normalized distance between two binary sequences. A number of distance measurement algorithms can be employed in the proposed scheme, such as the Hamming distance, Euclidean distance, etc. Note that in this chapter, Euclidean distance algorithm is used to carry out our simulation experiments.

5.4.2 Dirichlet-Based Reputation Model

In the system model, the control center takes the responsibility to monitor and assess the operating behaviors of each LA, and to determine whether an LA has been compromised or not, by analyzing the historical observations. As is known, Bayesian statistics is able to measure the uncertainty of a decision, as well as to offer future knowledge of such a decision given a line of historical observations. Motivated by this, in our work, we use a Bayesian statistics methodology to assist the control center in making right determinations of whether an LA has been compromised or not, as well as providing the control center with the knowledge of each LA's most probable behaviors in the near future. Specifically, as for the statistics methodology, Beta distribution is a valuable technique to determine whether a determination is right or not, while the Dirichlet distribution is able to determine in what degree a decision is correct [24]. Therefore, in this chapter, we consider employing a Dirichlet-based probabilistic model to obtain an accurate assessment of each LA's behaviors and hence an accurate decision.

The Dirichlet distribution is established on original beliefs of an undiscovered event described by a prior distribution. These initial beliefs along with a line of historical observations can be described by a posterior distribution. The posterior distribution is best suited to build our reputation model, because it is required that the reputation values are to be updated on the basis of historical observations. Let X be a discrete random variable which denotes the compliance level ρ of the measurement data for an LA. X can take values in the set $X = \{x_1, x_2, \ldots, x_k\}$, where $x_i \in [0, 1]$ and $x_{i+1} > x_i$ $(i = 1, \ldots, k)$. Generally, we have $x_1 = 0$, and $x_k = 1$. Let $\mathbf{p} = [p_1, p_2, \ldots, p_k]$ with $\sum_{i=1}^{k} p_i = 1$ be the probability distribution of random variable X, that is $p\{X = x_i\} = p_i$. In addition, let $\boldsymbol{\zeta} = [\zeta_1, \zeta_2, \ldots, \zeta_k]$ be the respective initial beliefs of each possible values of X. In this way, we model $\mathbf{p}$ with a posterior Dirichlet distribution, which is given by

$$f(\mathbf{p}|\boldsymbol{\zeta}) = \frac{1}{B(\boldsymbol{\zeta})} \prod_{i=1}^{k} p_i^{\zeta_i - 1} = \frac{\Gamma(\zeta_0)}{\prod_{i=1}^{k} \Gamma(\zeta_i)} \prod_{i=1}^{k} p_i^{\zeta_i - 1}, \tag{5.9}$$

where $\Gamma(\cdot)$ is the Gamma function, while $B(\cdot)$ is the Beta function. The total number of beliefs $\zeta_0 = \sum_{i=1}^{k} \zeta_i$. Given the historical statistics $\boldsymbol{\zeta}$, the expectation of the probability of $\mathbf{X}$ taking a value of x_i is shown below

$$E(p_i|\boldsymbol{\zeta}) = \frac{\zeta_i}{\zeta_0}. \tag{5.10}$$

Let $p_i^j(t)$ be the probability where LA_j behaves with a compliance level of x_i at time instant t, where $\sum_{i=1}^{k} p_i^j(t) = 1$. The, we model $p_i^j(t)$ by employing a posterior Dirichlet distribution according to Eq. (5.9). Prior to that, we define a random variable $Y^j(t)$ which denotes the summed products of the grade and probability of each compliance level in $\mathbf{p}^j(t) = [p_1^j(t), p_2^j(t), \ldots, p_k^j(t)]$ for LA_j, that is

$$Y^j(t) = \boldsymbol{\omega}\mathbf{p}^j(t) = \sum_{i=1}^{k} \omega_i p_i^j(t), \tag{5.11}$$

where $\boldsymbol{\omega} = [\omega_1, \omega_2, \ldots, \omega_k]$ are the grades assigned to each compliance level, implying the diverse impacts of each compliance level on LA_j's overall operating performance. Such design can considerably improve the accuracy of the control center's decision-making.

In order to evaluate the overall performance of an LA's behavior, we define the *reputation level*. In specific, each LA's behavior can be represented by using various compliance levels. As a result, it is natural to define the reputation level of an LA by the graded expectation of each compliance level, which is shown below:

$$R^j(t) = E[Y^j(t)] = \sum_{i=1}^k \omega_i E[p_i^j(t)] = \frac{1}{\zeta_0^j(t)} \sum_{i=1}^k \omega_i \zeta_i^j(t), \qquad (5.12)$$

where $\zeta_i^j(t)$ is the cumulative historical observations of LA_j at time instant t with compliance level x_i. Then, the variance of $Y^j(t)$ is expressed as

$$\sigma^2[Y^j(t)] = \sum_{i=1}^k \sum_{l=1}^k \omega_i \omega_l cov[p_i^j(t), p_l^j(t)]. \qquad (5.13)$$

Also, the covariance of $p_i^j(t)$ and $p_l^j(t)$ is computed by

$$cov[p_i^j(t), p_l^j(t)] = \frac{-\zeta_i^j(t)\zeta_l^j(t)}{\left(\zeta_0^j(t)\right)^2\left(\zeta_0^j(t)+1\right)}. \qquad (5.14)$$

5.4.3 Description of DDOA

In the proposed DDOA scheme, the control center, at the first step, trains the initial reputation levels of each LA given the collected historical observations, according to Algorithm 5.1. Although this algorithm is designed for the training phase and can be always finished offline, it is worth noting that the time complexity of this algorithm is $O(M \times N)$.

After finishing this training stage, the control center has the knowledge of the initial reputation levels of each LA. However, these initial reputation levels denote only the historical operating performance of each LA. Recall that smart grids are envisioned to support near real-time monitoring and control of the entire power grids. In this regard, consistent observations and assessments of each LA's behaviors are always demanded, in purpose of detecting whether an LA has been compromised or not.

In the detection stage, in order to update each LA's reputation level, a novelty algorithm including *reputation incentive mechanism* is proposed, the details of which can be seen in Algorithm 5.2. Specifically, given historical experiences, the control center defines two thresholds for the reputation levels, that is, H_s and H_m serving as the threats detection criteria, where H_s denotes a threshold for suspicious compliance level, while H_m denotes a threshold for malicious compliance level. In real-world scenarios, occurrence of system faults in smart grids is sometimes

---

Algorithm 5.1 The reputation level training

1: **procedure** DIRICHLET-BASED REPUTATION LEVEL TRAINING
2:     **for** $j = 1$ to M, the control center **do**   ▷ M is the total number of LAs
3:         1). Extracts N pieces of reported measurement data from LA_j;
4:         2). Computes the compliance level of each piece of measurement data $\rho^j(t), t \in [1, N]$
    as per Eq. (5.8);
5:         **for** $t = 1$ to N **do**
6:             **for** $i = 1$ to k **do**
7:                 **if** $\rho^j(t) = x_i$ **then**
8:                     $\zeta_i^j(t) \leftarrow \zeta_i^j(t-1) + 1$;
9:                     **break**;
10:                **else**
11:                    $\zeta_i^j(t) \leftarrow \zeta_i^j(t-1)$;
12:                **end if**
13:            **end for**
14:            a). $\zeta_0^j(t) = \sum_{i=1}^{k} \zeta_i^j(t)$;
15:            b). Computes the reputation level of LA_j by
16:                $R^j(t) = \frac{1}{\zeta_0^j(t)} \sum_{i=1}^{k} \omega_i \zeta_i^j(t)$.
17:        **end for**
18:    **end for**
19: **end procedure**

---

inevitable and, as a result, leads to wide fluctuations of state variables. Such events can, most possibly, affect and reduce both the compliance level as well as the reputation level. If a singular detection threshold is employed, we may probably have high false positives. On the contrast, two levels of detection threshold can, to a large extent, tolerate such system faults. As a result, the false positive rate can be significantly reduced, and the detection rate can be considerably improved.

By comparing the real-time reputation level to the two given thresholds, an LA can be categorized into one of the following three groups, that is, normal, suspicious, and malicious group.

- *normal group* ($\mathbb{N}$): LAs resided in the normal group are considered as benign. In this case, no further actions need to be taken.
- *suspicious group* ($\mathbb{S}$): LAs fell into the suspicious group will trigger the reputation incentive mechanism, which is designed to adapt the frequency of monitoring and assessment, as well as the grades assigned to each compliance level.
- *malicious group* ($\mathbb{M}$): LAs belong to the malicious group are considered as malicious. That is to say, they are much likely to be compromised by opportunistic FcDI attackers.

It is worth noting that, from a social perspective, one builds up a good social reputation by spending a significant amount of time persistently behaving in "good." However, only a few "bad" behaviors can lead to severe negative impacts on an individual's social reputation, which may possibly suffers a quick decrease in this case [18]. Motivated by this, we design a reputation incentive mechanism for LAs

---

Algorithm 5.2 DDOA algorithm

1: **procedure** REPUTATION UPDATING AND INTRUSION DETECTION
2:     **Initialization:**
3:     T_{max}, T_{min}, T_S, T_W, $H_s > H_m$, $N_{count} = 0$,
4:     $\mu_1 > \mu_2 > \cdots > \mu_k$, $T_1 = T_2 = \cdots = T_M = T_{max}$,
5:     $\omega_1 = \overline{\omega}_1$, $\omega_2 = \overline{\omega}_2$, $\ldots$, $\omega_k = \overline{\omega_k}$
6:     **for** $j = 1$ to M, control center **do** with a frequency of $1/T_j$
7:         1). Input: $\rho^j(t)$, $R^j(t-1)$, ω_1^j, ω_2^j, $\ldots$, ω_k^j
8:         2). Classification:

9:     $$LA_j \in \begin{cases} \mathbb{N}, & \text{if } R^j(t-1) > H_s \\ \mathbb{S}, & \text{if } H_s \geq R^j(t-1) \geq H_m \\ \mathbb{M}, & \text{if } R^j(t-1) < H_m \end{cases}$$

10:         3). Judgement:
11:     **switch** LA_j **do**
12:         **case:** $LA_j \in \mathbb{N}$
13:             a). LA_j is benign;
14:             b). $\omega_k^j \leftarrow \min\{\omega_k^j e^{\mu_k}, 1\}$;
15:             c). $\omega_i^j \leftarrow \overline{\omega}_i, \forall i = 1, 2, \ldots, k-1$;
16:             d). $T_j \leftarrow T_{max}$;
17:         **case:** $LA_j \in \mathbb{S}$
18:             $T_j \leftarrow \max\{T_j/2, T_{min}\}$;
19:             **if** $\rho^j(t) = x_k$ **then** $\triangleright x_k = 1$
20:                 $\omega_k^j \leftarrow \min\{\omega_k^j e^{\mu_k}, 1\}$;
21:                 $T_{count} \leftarrow T_{count} + 1$;
22:                 **if** $T_{count} > T_S$ **then**
23:                     $T_j \leftarrow \min\{T_j * 2, T_{max}\}$;
24:                     $T_{count} \leftarrow 0$;
25:                 **end if**
26:             **else**
27:                 $\omega_k^j \leftarrow \omega_k^j e^{-\mu_k}$;
28:                 **if** $\rho^j(t) = x_i$ $(i \neq k)$ **then**
29:                     $\omega_i^j \leftarrow \omega_i^j e^{-\mu_i}$;
30:                 **end if**
31:                 $T_{count} \leftarrow 0$;
32:             **end if**
33:         **case:** $LA_j \in \mathbb{M}$
34:             LA_j is compromised.
35:         4). Updates ζ_i^j for $i = 1, 2, \ldots, k$ with reference to Algorithm 5.1.
36:         5). Determines $R^j(t)$ using Eq. (5.12) with observation window
37:             T_W.
38:     **end for**
39: **end procedure**

---

resided in the suspicious group, in purpose of achieving an adaptive assessment of their operating performance. Specifically, in this mechanism, the grade ω_k is increased responding to an input value of $x^j(t) = x_k$, the full compliance level, and both the ω_k and ω_i are decreased in response to an input value of $x^j(t) = x_i$, $i \neq k$. Furthermore, in cases where LA_j falls in the suspicious group $\mathbb{S}$, the control center increases the frequency of monitoring and assessment over LA_j by two-fold, i.e., $T_j \leftarrow T_j/2$, in purpose of paying closer attention to this LA. In addition, under normal conditions, the control center monitors the operating performance of all the LAs with a constant period T_{max}. In cases where the control center observes that LA_j operates in good performance with all full compliance levels in a safe observation time period T_S, we reduce the frequency of monitoring and assessment by a half, i.e., $T_j \leftarrow T_j * 2$. In particular, as for cases where any LA recovers from the suspicious group $\mathbb{S}$ back to the normal group $\mathbb{N}$, the frequency of monitoring and assessment as well as all the grades, with the exception of ω_k, will be reset to the initial settings. It is worth noting that in this chapter, we monitor and assess each LA's operating behaviors within a long period, which is defined by an observation window T_W. In other words, control center only needs to assess LA's behavior within a time period of $[t - T_W, t]$. Note that the time complexity of Algorithm 5.2 is $O(M)$.

The purpose of designing such an incentive mechanism falls into the following two aspects. On the one hand, it can encourage non-malicious LAs (those who belong to the normal or suspicious group possibly due to system noises) to stick to performing good behaviors, in purpose of increasing or keeping their reputation levels. On the other hand, this mechanism can also quickly decrease the reputation level of a suspicious LA, if non-compliance behaviors are observed, which can respond fast to "bad" behaviors.

5.5   Performance Evaluation

In this section, a line of experiments have been conducted to evaluate the effectiveness of the proposed DDOA scheme. Specifically, first we perform Time Step Simulation experiments by employing the PowerWorld simulator [25], in order to collect a large amount of real-time measurement data from the IEEE 39-bus power testing system. After that, a set of simulations are carried out in MATLAB 2014b in purpose of analyzing these collected data.

5.5.1   Data Collection in PowerWorld

We utilize the IEEE 39-bus power system as our testing system. As shown in Fig. 5.3), the system is geographically divided into m areas, which is called LAs. Here, the m is set as 6 in our simulation. In PowerWorld, we make use of Time Step

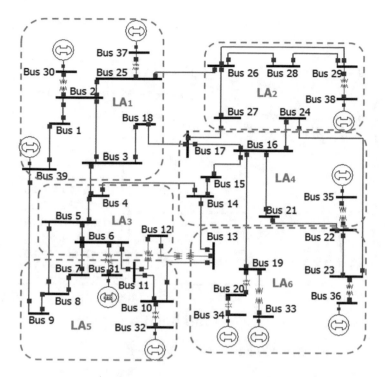

Fig. 5.3 IEEE 39-bus power system with example flocking areas

Simulation to collect massive real-time data for around 20,000 minutes, including power generations of each generator **G**, power loads of each bus **L**, and line power flows of each transmission line **F**, etc. The first 1500 min were used for the training phase and the rest for the detection phase.

Virtual data are inserted randomly into the collected data to simulate the behavior of LAs under different scalar ϵ and system noise P_n.

5.5.2   Data Analytics in MATLAB

Using the novelty reputation level training algorithm, the reputation levels are analyzed through utilizing the collected data. In the training stage, we first evaluate the effects of different ϵ and system noise P_n. The reputation levels with respect to ϵ along the training period are plotted in Fig. 5.4. It can be observed that the reputation level converges to a constant value as time progresses, and the higher the ϵ, the higher the reputation. This is because, as described in Sect. 5.2.2, a higher ϵ represents a lower probability of attack, therefore resulting in a higher reputation level. The reputation levels under different system noise P_n during

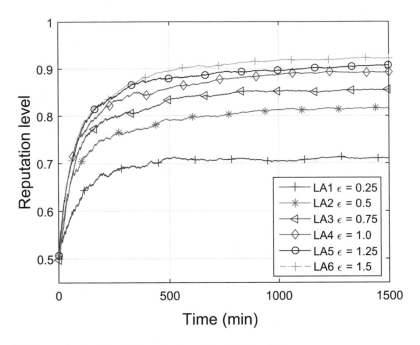

Fig. 5.4 Reputation level vs. different values of ϵ with $P_n = 0.1$

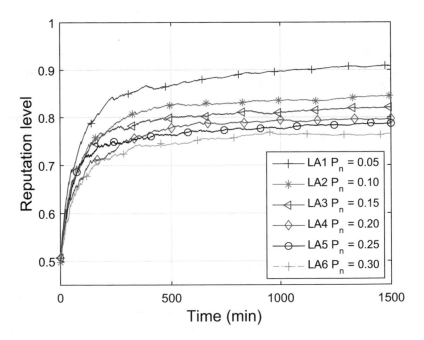

Fig. 5.5 Reputation level vs. different values of P_n with $\epsilon = 0.75$

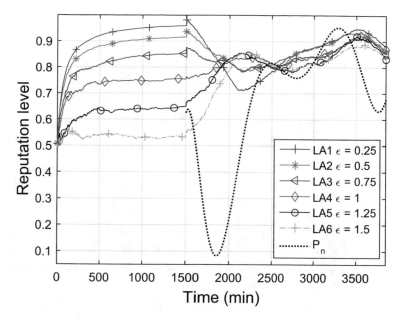

Fig. 5.6 Reputation level vs. different values of ϵ with daily dynamic P_n

training are plotted in Fig. 5.5. Similarly, the lower system noise leads to lower attack probability, while the reputation level gradually converges to a constant value.

Moreover, we consider that opportunistic attackers can adjust their attack probabilities according to the system noise, to demonstrate this, we profile the daily system noise level based on real-time daily load pattern in Fig. 5.6. Chertkov et al. have demonstrated a significant correlation between system noise and load pattern in [26].

In this case, the reputation level versus system noise level under different ϵ is also displayed. It can be seen from this figure that the system noise fluctuates conversely with the reputation level, owing to the same reason (that is, system noise has opposite influences on the reputation level).

In this case, we also show the relationship between the credit rating under different *epsilon* and the system noise level. As can be seen from the figure, the level of reputation fluctuates in the opposite direction to the level of system noise for the same reason (i.e., the effect of system noise on reputation is opposite).

In the detection stage, for the purpose of demonstrating the effectiveness of our proposed scheme, two scenarios are considered. In the first scenario (see Fig. 5.7), LA_2 is assumed to be compromised at the instant of 10,000 min, and is compromised by an attacker. It can be seen that, after LA_2 is compromised, the reputation level of LA_2 drops slightly from the normal group to the suspicious group threshold H_S. With the proposed reputation incentive mechanism, one is considered as suspicious once the reputation level drops below H_S, and the reputation level decreases rapidly to the malicious group threshold H_M if there are continuous non-

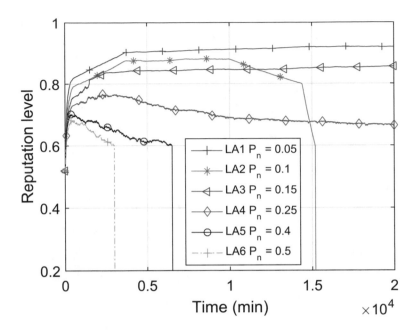

Fig. 5.7 Reputation level with an aggregative attacker with $\epsilon = 0.75$

compliance behaviors. Therefore, the compromised LA_2 has been detected. On the contrary, LA_5 and LA_6 are set to be compromised from the beginning. A significant difference is that the system noise of LA_6 is higher than LA_5, and the reputation level of LA_6 declines faster than LA_5.

In the second scenario (see Fig. 5.8), a temporal system fault is inserted to LA_3 at the instant of 10,000 min for the purpose of highlighting the difference between attackers and system faults. It can be seen that due to the system fault, the reputation level of LA_3 first drops from the normal group to the suspicious group. Moreover, it is worth noting that the decrease rate of LA_3 in the normal group is lower than that in the suspicious group. The reason is that the proposed reputation incentive mechanism can adaptively change the decline rate accordingly. Then, the reputation level gradually recovers and, finally, converges to a stable level. Obviously, the system fault will not change the behavior of the LA, and although the reputation decreases within a short period of time, our scheme can restore the reputation.

Finally, the relationship between the detection rate and the length of the observation window T_W is shown in Fig. 5.9. It can be seen that within a certain period (e.g., [2000, 4000]), the detection rate will improve with the addition of the observation window length. The reason is that a longer observation window can provide additional evidence to detect the hidden attackers. Compared with conservative attackers (with $\epsilon > 1$), it is faster to detect the aggressive attackers (with $\epsilon < 1$) using our proposed scheme.

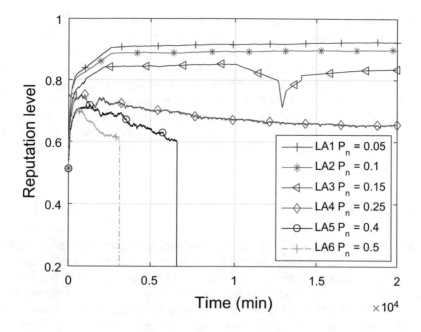

Fig. 5.8 Reputation level with an inserted temporal system fault with $\epsilon = 0.75$

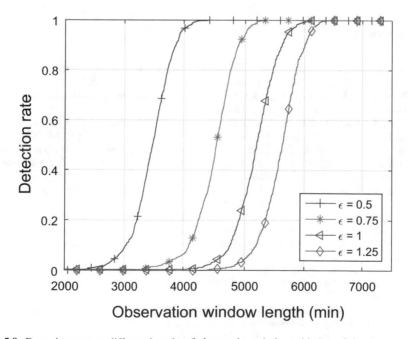

Fig. 5.9 Detection rate vs. different lengths of observation window with $P_n = 0.1$

In conclusion, we have shown that a potential class of opportunistic attackers in smart grids can adjust their attack probabilities according to the dynamic system noise level P_n, and the proposed DDOA scheme can effectively detect and identify these opportunistic attackers (e.g., state-sponsored actors). We have also shown that our proposed scheme dose not treat occasional system faults as attacks due to the two specified thresholds H_s and H_m. Moreover, we have demonstrated that a high detection rate can be obtained with long observation windows in the proposed scheme. Thus, our scheme is an effective and promising solution for detecting opportunistic attackers in smart grid cyber-physical systems.

5.6 Summary

In this chapter, a three-tier hierarchical framework for future smart grids has been introduced. In addition, the importance of resilience against financially motivated opportunistic attackers (seeking to manipulate smart electricity prices) has also been emphasized. For the purpose of preventing opportunistic attacks, a Dirichlet-based detection scheme (DDOA) has been proposed to detect and identify attackers, and the effectiveness of the proposed scheme has been demonstrated using simulations of extensive real-time data collected from the IEEE 39-bus power testing system.

References

1. Bao, H., Lu, R., Li, B., & Deng, R. (2016). BLITHE: Behavior rule based insider threat detection for smart grid. *IEEE Internet of Things Journal, 3*(2), 190–205.
2. Yuan, Y. L., Li, Z. Y., & Ren, K. (2011). Modeling load redistribution attacks in power systems. *IEEE Transactions on Smart Grid, 2*(2), 382–390.
3. Deng, R., Xiao, G., & Lu, R. (2017). Defending against false data injection attacks on power system state estimation. *IEEE Transactions on Industrial Informatics, 13*(1), 98–207.
4. Vukovic, O., & Dan, G. (2014). Security of fully distributed power system state estimation: Detection and mitigation of data integrity attacks. *IEEE Journal on Selected Areas in Communications, 32*(7), 1500–1508.
5. Kekatos, V., & Giannakis, G. B. (2013). Distributed robust power system state estimation. *IEEE Transactions on Power Systems, 28*(2), 1617–1626.
6. Iwamoto, S., Kusano, M., & Quintana, V. H. (1989). Hierarchical state estimation using a fast rectangular-coordinate method. *IEEE Transactions on Power Systems, 4*(3), 870–880.
7. ICS-CERT. (2015, November/December). *NCCIC/ICS-CERT Monitor November–December 2015.*
8. Zetter, K. (2015, August). *A cyberattack has caused confirmed physical damage for the second time ever* [Online]. Available: https://www.wired.com/2015/01/german-steel-mill-hack-destruction/
9. Pagliery, J. (2015, October). *ISIS Is Attacking the U.S. Energy Grid*. http://money.cnn.com/2015/10/15/technology/isis-energy-grid/. Accessed 28 June 2018.
10. Bi, S. Z., & Zhang, Y. J. (2013). False-data injection attack to control real-time price in electricity market. In *Proceedings of the IEEE GLOBECOM, Atlanta, GA, USA, December 9–13* (pp. 772–777).

11. Xie, L., Mo, Y., & Sinopoli, B. (2010). False data injection attacks in electricity markets. In *Proceedings of the First IEEE International Conference on Smart Grid Communications (SmartGridComm), Gaithersburg, MD, USA, October 4–6*, 2010 (pp. 226–231).
12. Tan, R., Krishna, V. B., Yau, D. K., & Kalbarczyk, Z. (2015). Integrity attacks on real-time pricing in electric power grids. *ACM Transactions on Information and System Security, 18*(2), 5.
13. Jia, L., Thomas, R. J., & Tong, L. (2012). Impacts of malicious data on real-time price of electricity market operations. In *Proceedings of the 45th Hawaii International Conference on System Sciences, Maui, HI, USA, January 04–07*, 2012 (pp. 1907–1914).
14. Kosut, O., Jia, L. Y., Thomas, R. J., & Tong, L. (2011). Malicious data attacks on the smart grid. *IEEE Transactions on Smart Grid, 2*(4), 645–658.
15. Esmalifalak, M., Shi, G., Han, Z., & Song, L. (2013). Bad data injection attack and defense in electricity market using game theory study. *IEEE Transactions on Smart Grid, 4*(1), 160–169.
16. Esmalifalak, M., Han, Z., & Song, L. (2012). Effect of stealthy bad data injection on network congestion in market based power system. In *Proceedings of the 2012 IEEE Wireless Communications and Networking Conference (WCNC), Paris, France, April 1–4*, 2012 (pp. 2468–2472)
17. Lu, R. X., Liang, X. H., Li, X., Lin, X. D., & Shen, X. M. (2012). EPPA: An efficient and privacy-preserving aggregation scheme for secure smart grid communications. *IEEE Transactions on Parallel Distributed Systems, 23*(9), 1621–1631.
18. Sun, Y. L., Han, Z., Yu, W., & Liu, K. R. (2006). A trust evaluation framework in distributed networks: Vulnerability analysis and defense against attacks. In *Proceedings of the 25th IEEE INFOCOM, Barcelona, Catalunya, Spain, April 23–29*, 2006 (pp. 1–13).
19. Li, S., Yilmaz, Y., & Wang, X. (2015). Quickest detection of false data injection attack in wide-area smart grids. *IEEE Transactions on Smart Grid, 6*(6), 2725–2735.
20. Ma, J., Liu, Y., Song, L., & Han, Z. (2015). Multiact dynamic game strategy for jamming attack in electricity market. *IEEE Transactions on Smart Grid, 6*(5), 2273–2282.
21. Mitchell, R., & Chen, I.-R. (2015). Behavior rule specification-based intrusion detection for safety critical medical cyber physical systems. *IEEE Transactions on Dependable Secure Computing, 12*(1), 16–30.
22. Ott, A. L. (2003). Experience with PJM market operation, system design, and implementation. *IEEE Transactions on Power Systems, 18*(2), 528–534.
23. Schweppe, F. C., Wildes, J., & Rom, D. B. (1970). Power system static-state estimation, parts I, II, and III. *IEEE Transactions on Power Apparatus and Systems, 89*(1), 120–135.
24. Johnson, N. L., Kotz, S., & Balakrishnan, N. (2002). *Continuous multivariate distributions, volume 1, Models and applications* (Vol. 59). New York: Wiley.
25. PowerWorld (2018). https://www.powerworld.com. Accessed 28 June 2018.
26. Chertkov, M., Pan, F., & Stepanov, M. G. (2011). Predicting failures in power grids: The case of static overloads. *IEEE Transactions on Smart Grid, 2*(1), 162–172.

Chapter 6
PFDD: On Feasibility and Limitations of Detecting FmDI Attacks Using D-FACTS

As reported by recent studies that D-FACTS (distributed flexible AC transmission system) devices is able to enable **Proactively** **FmDI** attack **Detection** on smart grids, which we term as the PFDD approach. However, few literature has focused on the systematic analysis of the feasibility and limitations of this approach. To meet this gap, in this chapter we pioneer to investigate the feasibility and limitations by employing PFDD to contain FmDI attacks on smart grids, by considering single-bus, uncoordinated multiple-bus, and coordinated multiple-bus FmDI attacks, respectively. It is proved that PFDD can detect all these three classes of FmDI attacks targeted on buses or super-buses with degrees larger than 1, as long as the deployment of D-FACTS devices covers branches at least containing a spanning tree of the grid graph. The minimum efforts required for activating D-FACTS devices to detect each class of FmDI attacks are respectively evaluated. In addition, the limitations of this approach are also discussed, and it is strictly proved that PFDD is not able to detect FmDI attacks targeted on buses or super-buses with degrees equalling 1.

6.1 Introduction

Recent years have witnessed a new threat landscape in power grids, among which, the high-profile FmDI attacks have drawn intensive research interests from both power and security communities [1–8]. The information of power grid topology and line configurations is critical to a successful FmDI attack. The deep integration of ICTs into the power grids and the widespread powerful hacking tools [9] have considerably prompted the information harvesting over power girds. As mentioned earlier in Sect. 2.3, various channels can be accessed by FmDI attackers to collect useful information of power grids without any authorization.

© Springer Nature Switzerland AG 2020   123
B. Li et al., *Detection of False Data Injection Attacks in Smart Grid
Cyber-Physical Systems*, Wireless Networks,
https://doi.org/10.1007/978-3-030-58672-0_6

Equipped with these valuable knowledge of power girds, FmDI attackers are able to construct desired attack vectors which can easily bypass the conventional false data detection (FDD) defenses based on power grid state estimations [4, 10, 11], which we call passive defense approaches. In this case, many of the existing passive defenses may no longer be feasible. Some studies have recently shown that it is possible to achieve proactive FDD in power grids by utilizing D-FACTS devices—we call the PFDD approach [12–14]. Morrow et al. are presumably the pioneers to develop the idea of achieving topology perturbation by using D-FACTS devices, in purpose of detecting bad data (caused by either malicious injection or system faults) in power grids [12]. Following this chapter, Rahman et al. investigated the moving target defense approaches to harden the security of the power system state estimation, one of which is to perturb the power line admittance by using D-FACTS devices [13]. More recently, Tian et al. designed a hidden moving target defense that is able to alter line susceptance values while maintaining the power flows, to avoid alerting the attackers who can compute the state estimation residuals [14].

Despite the above-mentioned encouraging developments, some important issues pertaining to the PFDD approach remain largely open, e.g., how much D-FACTS devices and where to deploy them across the power system can help with PFDD, especially when the FmDI attack strategies are evolving rapidly and appearing in a more sophisticated and coordinated mode [10]. In this chapter, our goal is to systematically explore the feasibility and potential limitations by utilizing the PFDD approach to identify FmDI attacks in smart grids. Three categories of FmDI attacks are, respectively, considered, namely single-bus, uncoordinated multiple-bus, and coordinated multiple-bus FmDI attacks. We adopt an arguably more realistic assumption that the adversaries can falsify the measurement data but cannot compute the state estimation residuals by themselves on real-time basis. It is based on the consensus that attackers are usually equipped with limited capabilities, unable to obtain the real-time global knowledge and measurement data of the entire power grid [3].

The main contributions of this chapter are fourfold:

- First, we design a framework to detect FmDI attacks on smart grids by utilizing the PFDD approach. Prior to that, the rationale for designing this framework is also presented.
- Second, we study on the feasibility of utilizing the PFDD approach to detect the aforementioned three categories of FmDI attacks on smart grids. We prove that this approach is able to identify the existence of all these FmDI attacks that targeted on power grid buses or super-buses having degrees larger than 1, as long as the deployment of D-FACTS devices covers branches at least forming a spanning tree of the grid graph.
- Third, we are able to derive the minimum effort profiles that demanded for activating D-FACTS devices (in order to identify FmDI attacks) versus the injected offsets on the system states, respectively, for each class of FmDI attacks. The power system defenders can, then, make appropriate efforts to contain FmDI attacks by making use of these valuable profiles.

- Last, the potential limitations by using the PFDD approach are also investigated. It can be strictly proved that the PFDD approach is incapable of detecting FmDI attacks targeted on power grid buses or super-buses having degrees 1.

The remaining parts of this chapter are organized as follows. Section 6.2 presents an overview of the D-FACTS devices, and details of our system model and the adversary model. In Sect. 6.3, we elaborate on the framework of the PFDD approach and the feasibility investigations, followed by the explorations on its limitations in Sect. 6.4. Section 6.5 closes this chapter with the conclusion.

6.2 Overview and Models

6.2.1 Overview of D-FACTS Devices

In power grids, the flexible AC transmission system (FACTS) devices have been proven as a feasible solution to control power flows in the transmission and distribution system by altering the impedance of the power lines or changing the phase angle of the voltage applied across the lines [15]. However, high costs and reliability concerns have limited the deployments of FACTS devices. A distributed solution of FACTS, named D-FACTS, has therefore emerged and got widely accepted because of its smaller scale, lower costs, and better performance compared to FACTS devices [15]. It can be reasonably expected that D-FACTS devices will be widely deployed across smart grids in the near future due to its increasing capabilities and decreasing installation costs [14, 16].

Representatives of D-FACTS devices include distributed static series compensator (DSSC), distributed series reactor (DSR), and synchronous voltage source (SVS) [17]. These D-FACTS devices can be used to support a myriad of applications such as contingency response, loop flow control, phase balancing, transient stability response, renewable energy transfers, etc. In this chapter, we employ D-FACTS devices with the expectation to achieve adaptable power grid configurations, which ultimately allows proactive detection of FmDI attacks.

6.2.2 System Model

In our system model, we mainly take into account the DC power flow based state estimation along with the conventional bad data detection (BDD) procedure, as shown in Fig. 6.1. Note that though we mainly focus on DC power flow model for its suitability and simplicity, we also extend our analysis to AC power flow model in Sect. 6.3.2.2 to show the universality of our studies.

The basic concepts of state estimation are introduced in Sect. 2.1. In this chapter, we provide more details of how the H matrix is constructed.

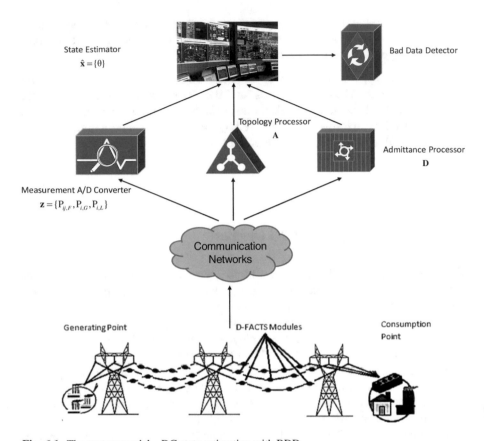

Fig. 6.1 The system model—DC state estimation with BDD

6.2.2.1 H Matrix Construction

Let $\mathbf{A} \in \mathbb{R}^{l \times n}$ denote the branch-bus connection matrix, where l is the total number of branches (power lines) in a power grid, and n is the total number of buses that is the same as the number of system states in DC flow based state estimation. The entries of $\mathbf{A}$ are given by

$$a_{ki} = \begin{cases} 1, & \text{if branch } k \text{ starts at bus } i \\ -1, & \text{if branch } k \text{ ends at bus } i \\ 0, & \text{otherwise}, \end{cases} \tag{6.1}$$

where $k \in \mathcal{L} = \{1, 2, \cdots, l\}$ and $i \in \mathcal{N} = \{1, 2, \cdots, n\}$. In addition, let $\mathbf{D} \in \mathbb{R}^{l \times l}$ denote a diagonal matrix with its diagonal entries being the admittance (negative susceptance in DC power flow model) values of branches. Then the Jacobian matrix $\mathbf{H}$ is constructed as [18]

$$H = \begin{bmatrix} A^{\mathsf{T}} DA \\ DA \\ -DA \end{bmatrix}. \tag{6.2}$$

6.2.3   Adversary Model

As mentioned in Sect. 2.3 that, to launch a successful FmDI attack, the knowledge of H matrix is demanded for the attackers. This is also described in Lemma 6.1.

Lemma 6.1 ([3]) *Suppose the original measurement data vector z can bypass the bad data detector. The fabricated malicious measurement data vector $z_a = z + a$ can also bypass the bad data detector if a is a linear combination of the column vectors of H (i.e., $a = Hc$).*

For attackers without the knowledge of H matrix, it is hard for them to select an attack vector a that can, fortunately, happen to lead to a success in passing the BDD test. In contrast, when it comes to knowledgeable attackers, they can choose any non-zero arbitrary vector c and construct an attack vector by $a = Hc$ if they have sufficient knowledge of the H matrix. As aforementioned, the recent years have seen the growing knowledge of the attackers; therefore, the existence of knowledgeable attackers is increasingly becoming a challenge that we must carefully handle. In this chapter, we take into account three classes of FmDI attacks in terms of the attackers' capabilities as well as their knowledge levels of the H matrix, including

- Single-bus FmDI attacks: The attackers are usually equipped with a weak attack capability and basic knowledge level of H matrix (i.e., the susceptance information of only one bus's incident branches). This class of FmDI attacks can only be mounted on a specific single-bus, e.g., $c_i = \theta_a$ for $i \in N$ and $c_j = 0$ for $\forall j \in N \setminus i$, where θ_a is a constant number.
- Uncoordinated multiple-bus FmDI attacks: The attackers are with intermediate attack capability and medium knowledge level of H matrix (i.e., the susceptance information of several buses' incident branches). This class of FmDI attacks can be concurrently but independently mounted on multiple power grid buses in an uncoordinated mode, for example, $c = (\underbrace{0, \theta_{a1}, 0, 0, \theta_{a2}, \theta_{a3}, 0, \cdots}_{n})^{\mathsf{T}}$, where
  θ_{a1}, θ_{a2}, and θ_{a3} are distinct constant numbers of voltage phase angle.
- Coordinated multiple-bus FmDI attacks (also termed as super-bus FmDI attacks [10]): The attackers are usually equipped with a strong attack capability and expert knowledge level of the H matrix, i.e., the susceptance information of a super-bus. This class of FmDI attacks can be concurrently carried out on multiple buses in a coordinated mode, for example, $c = (\underbrace{\theta_a, \theta_a, 0, 0, \theta_a, 0, \theta_a, \cdots}_{n})^{\mathsf{T}}$.

6.3 The Feasibility of PFDD

In this section, we explore the feasibility of utilizing the PFDD approach to identify FmDI attacks on smart grids. Prior to delving into the detailed findings, we give the definition of the degree of a bus in power grids.

Definition 6.1 Given the graph of a power grid topology, the degree of a bus is defined as the number of connections (branches) it has to other buses. Similarly, the degree of a super-bus is defined as the number of connections (branches) the super-bus has to other buses.

Note that there is usually no isolated buses with degrees 0 in real-world power grids. In this chapter, we consider cases where the degree of any bus in a power grid is no less than 1. Our discussions contain two parts. In the first part, we show that the PFDD approach can detect FmDI attacks targeted on those buses with degrees larger than 1, of which the details will be presented later in this section. In the second part, we show that for buses with degrees 1, the PFDD approach cannot detect FmDI attacks targeted on them; the details will be presented in Sect. 6.4. Note that hereafter in this section, all buses we are talking about are with degrees strictly larger than 1.

6.3.1 The Framework for PFDD Approach

The PFDD approach fulfills FDD by (1) proactively activating the D-FACTS devices deployed on the transmission lines (branches), (2) updating state estimation parameters, and (3) conducting BDD process. Activating D-FACTS devices proactively changes the system configuration information, therefore affecting the state estimation; however, the attackers are incapable of following such configuration changes in a very short time. This builds up the premises for deploying the PFDD approach in smart grids to detect FmDI attacks. Note that, unlike most of the existing FDD approaches, PFDD is applied proactively regardless of whether anomaly is observed/sensed.

We develop a framework for the PFDD approach as shown in Algorithm 6.1. The rationale behind is discussed below. Assume that when D-FACTS devices are triggered, the values of line admittance are changed by

$$\mathbf{D}' = \mathbf{D} + \Delta\mathbf{D}, \tag{6.3}$$

where $\Delta\mathbf{D}$ is a matrix with the same dimensions to $\mathbf{D}$, encompassing the variations of line admittance when D-FACTS devices are triggered. Correspondingly, the measurement Jacobian matrix is altered by

$$\mathbf{H}' = \begin{bmatrix} \mathbf{A}^\mathsf{T}\mathbf{D}'\mathbf{A} \\ \mathbf{D}'\mathbf{A} \\ -\mathbf{D}'\mathbf{A} \end{bmatrix} = \begin{bmatrix} \mathbf{A}^\mathsf{T}(\mathbf{D}+\Delta\mathbf{D})\mathbf{A} \\ (\mathbf{D}+\Delta\mathbf{D})\mathbf{A} \\ -(\mathbf{D}+\Delta\mathbf{D})\mathbf{A} \end{bmatrix} = \mathbf{H} + \Delta\mathbf{H}, \qquad (6.4)$$

where

$$\Delta\mathbf{H} = \begin{bmatrix} \mathbf{A}^\mathsf{T}\Delta\mathbf{D}\mathbf{A} \\ \Delta\mathbf{D}\mathbf{A} \\ -\Delta\mathbf{D}\mathbf{A} \end{bmatrix}. \qquad (6.5)$$

By conducting state estimation when false data is injected, the Frobenius norm of the normalized measurement residuals is updated by

$$\begin{aligned} \|\overline{\boldsymbol{\gamma}}'_a\| &= \|\sqrt{\mathbf{W}^{-1}}(\mathbf{z}'_a - \mathbf{H}'\hat{\mathbf{x}}'_a)\| \\ &= \|\sqrt{\mathbf{W}^{-1}}[\mathbf{z}' + \mathbf{a} - \mathbf{H}'(\hat{\mathbf{x}}' + \Delta\mathbf{x})]\| \\ &= \|\sqrt{\mathbf{W}^{-1}}(\underbrace{\mathbf{z}' - \mathbf{H}'\hat{\mathbf{x}}'}_{original} + \underbrace{\mathbf{a} - \mathbf{H}'\Delta\mathbf{x}}_{injected})\|, \end{aligned} \qquad (6.6)$$

where $\mathbf{z}'$ is the updated measurement vector after D-FACTS devices are triggered, $\hat{\mathbf{x}}'$ denotes the corresponding estimated system states, and $\Delta\mathbf{x}$ is the injected offsets on system states, respectively. We recall that the attackers are not able to immediately obtaining the full knowledge of the updated measurement Jacobian matrix $\mathbf{H}'$ right after the D-FACTS devices are triggered; hence, they still construct the attack vector by $\mathbf{a} = \mathbf{Hc}$ by making use of the $\mathbf{H}$. In this case, the reported falsified measurement data $\mathbf{z}'_a = \mathbf{z}' + \mathbf{Hc}$ can be easily identified as being abnormal. This is because that in most cases, vector $\sqrt{\mathbf{W}^{-1}}(\mathbf{a} - \mathbf{H}'\Delta\mathbf{x}) = \sqrt{\mathbf{W}^{-1}}(\mathbf{Hc} - \mathbf{H}'\Delta\mathbf{x})$, the *injected* part of Eq. (6.6), no longer equals $\mathbf{0}$. When the entry values of this vector are sufficiently large, it leads to $\|\overline{\boldsymbol{\gamma}}'_a\| > \tau$ and triggers the false data alarm.

Algorithm 6.1 The framework for PFDD approach

1: **procedure**
2:     1). Trigger the D-FACTS devices of interest;
3:     2). Update the $\mathbf{D}$ matrix by Eq. (6.3);
4:     3). Update the $\mathbf{H}$ matrix by Eq. (6.4);
5:     4). Conduct state estimation by Eq. (2.6) using the updated $\mathbf{D}'$ and $\mathbf{H}'$ matrices;
6:     5). Perform the BDD procedure by Eq. (2.38):
7:     **if** $\|\overline{\boldsymbol{\gamma}}'_a\| > \tau$ **then**
8:        **output:** FmDI attack is identified.
9:     **else**
10:       **output:** No FmDI attack is identified.
11:    **end if**
12: **end procedure**

6.3.2   Evaluation of the Minimum Efforts Demanded for D-FACTS Devices to Identify Effective FmDI Attacks

We start our discussions by making the following definitions.

Definition 6.2 An FmDI attack is defined as *effective FmDI attack*, if the injected false measurement data can eventually cause real alterations on the power flows, while an *ineffective FmDI attack*, though being able to inject falsified measurement data, cannot eventually cause real alterations on the power flows. Particularly, FmDI attacks, that the entries of the injected offsets c are not exceeding the tolerance thresholds of system state errors/faults, are regarded as an *ineffective FmDI attack*, because such small-value false measurement data cannot cause real alterations and, therefore, can be tolerated.

Note that there is a special case for the *ineffective FmDI attack* would be a coordinated multiple-bus FmDI attack on all the buses, i.e., $c = (\theta_a, \theta_a, \cdots, \theta_a)^{\mathsf{T}}$. In this case, though the attacker is able to successfully inject false measurement data $z_a = z + Hc$ and leading to $x_a = x + c$, it is not able to cause any alteration on the power flows. This is because, according to Eq. (6.8), the power flow is proportional to the voltage phase difference between buses. The attack injecting a same value of voltage phase angle to all buses, where no phase difference is made between any two buses; thus, no alteration on the power flows can be caused.

Although PFDD is theoretically feasible, it is still not clear enough in the power community whether proactively activating D-FACTS devices will contribute to potential hidden impacts or instability issues. To help provide necessary knowledge of making a consensus on this question, it is valuable to figure out the minimum efforts needed for using D-FACTS to identify *effective* FmDI attacks.

6.3.2.1   Optimization Problem Formulation Under DC Model

We define the required *efforts*, resulting from activating D-FACTS devices to identify the existence of *effective* FmDI attacks, as $\|\Delta D\|$, the Frobenius norm of the line admittance variations on all the branches. This optimization problem is to minimize the required *efforts* under the constraints of power flow balance laws and D-FACTS capabilities, which is given by

$$\min_{\Delta D} \quad \|\mathrm{diag}(\Delta D)\| \tag{6.7a}$$

$$\text{s.t.} \quad \|\overline{\gamma}'_a(\Delta D)\| > \tau \tag{6.7b}$$

$$d_k^{min} < \Delta d_{kk} < d_k^{max}, \ k \in \mathcal{L} \tag{6.7c}$$

$$P_{i,G} - P_{i,L} = \sum_{j \in \mathcal{N}_i} P'_{ij,F}, \ i, j \in \mathcal{N}, \tag{6.7d}$$

where the function diag($\cdot$) outputs a vector encompassing the diagonal entries of a square matrix. Since $\Delta \mathbf{D}$ is a diagonal matrix, $\|\Delta \mathbf{D}\|$ is equivalent to $\|\text{diag}(\Delta \mathbf{D})\|$. Δd_{kk} is the k-th entry of vector diag($\Delta \mathbf{D}$), and $\overline{\boldsymbol{\gamma}}_a'(\Delta \mathbf{D})$, a function of $\Delta \mathbf{D}$, represents the updated normalized estimation residual vector with false measurement data injected. d_k^{min} and d_k^{max} serve as the lower and upper bounds of Δd_{kk}, respectively, implying the range of admittance variations that D-FACTS devices implemented on the k-th branch can realize. $P_{i,G}$ and $P_{i,L}$, respectively, denote the power generation(s) and power load(s) at bus i. The neighbor buses of bus i are denoted by $\mathcal{N}_i$, and the updated power flow between buses i and j is denoted by $P_{ij,F}'$, that is, in DC model, given by

$$P_{ij,F}' = d_{kk}'(\theta_i' - \theta_j') = -b_{ij}'(\theta_i' - \theta_j'), \tag{6.8}$$

where b_{ij}' denotes the updated susceptance of branch (i, j), also denoted by the k-th branch (i.e., $b_{ij}' = -d_{kk}'$), and θ_i' and θ_j', respectively, denote the updated voltage phase angle on buses i and j.

As for those constraints in this optimization problem, formulas (6.7c) and (6.7d), respectively, describe the D-FACTS devices' capabilities and the optimal power flow balance requirements, respectively. Importantly, the formula (6.7b) describes the successful detection of FmDI attacks via the BDD process. The updated estimated system states with false data injected $\hat{\mathbf{x}}_a'$ are equivalent to the real updated system state vector added by the injected offset vector, which is given by

$$\hat{\mathbf{x}}_a' = \hat{\mathbf{x}}' + \Delta \mathbf{x}. \tag{6.9}$$

Also, according to Eq. (2.6), we have

$$\hat{\mathbf{x}}_a' = \mathbf{\Lambda}' \mathbf{z}_a' = \mathbf{\Lambda}'(\mathbf{z}' + \mathbf{a}) = \hat{\mathbf{x}}' + \mathbf{\Lambda}' \mathbf{a}. \tag{6.10}$$

Hence, $\Delta \mathbf{x}$ can be expressed as

$$\Delta \mathbf{x} = \mathbf{\Lambda}' \mathbf{a}. \tag{6.11}$$

In this way, constraint (6.7b) can be rewritten by

$$\begin{aligned}
\tau &< \|\overline{\boldsymbol{\gamma}}_a'(\Delta \mathbf{D})\| \\
&= \|\sqrt{\mathbf{W}^{-1}}(\mathbf{z}' - \mathbf{H}'\hat{\mathbf{x}}' + \mathbf{a} - \mathbf{H}'\Delta \mathbf{x})\| \\
&= \|\sqrt{\mathbf{W}^{-1}}(\mathbf{z}' - \mathbf{H}'\hat{\mathbf{x}}' + \mathbf{Hc} - \mathbf{H}'\mathbf{\Lambda}'\mathbf{Hc})\| \\
&= \|\sqrt{\mathbf{W}^{-1}}[\mathbf{z}' - \mathbf{H}'\hat{\mathbf{x}}' + (\mathbf{I} - \mathbf{H}'\mathbf{\Lambda}')\mathbf{Hc}]\| \\
&= \|\sqrt{\mathbf{W}^{-1}}\{\mathbf{z}' - (\mathbf{H} + \Delta \mathbf{H})\hat{\mathbf{x}}' + [\mathbf{I} - (\mathbf{H} + \Delta \mathbf{H})\mathbf{\Lambda}']\mathbf{Hc}\}\|.
\end{aligned} \tag{6.12}$$

Note that for clearance purpose, we will not further substitute $\Delta\mathbf{H}$ by $\Delta\mathbf{D}$, but it is worth noting that $\Delta\mathbf{D}$ well reflects the changes of $\Delta\mathbf{H}$ with reference to Eq. (6.5).

The formulated optimization problem and inequality shown in Eq. (6.12) permit us to assess the relations between the minimum $\|\text{diag}(\Delta\mathbf{D})\|$ versus the $\mathbf{c}$, and it is possible to obtain a general profile between the two, given a data measurement system with specific $\mathbf{A}$, $\mathbf{D}$, $\mathbf{W}$, and τ. At first glance, it seems that this relationship depends on real-time measurements $\mathbf{z}$ and system state vector $\mathbf{x}$. However, as per Sect. 6.2.2, we notice that $\|\sqrt{\mathbf{W}^{-1}}(\mathbf{z}' - (\mathbf{H}+\Delta\mathbf{H})\hat{\mathbf{x}}')\| < \tau$ holds all the time under normal circumstances. Therefore, it can be claimed that the entry values of vector $\sqrt{\mathbf{W}^{-1}}(\mathbf{z}' - (\mathbf{H}+\Delta\mathbf{H})\hat{\mathbf{x}}')$ should be sufficiently small, and thus can be reasonably ignored. Hence, with reference to Eq. (6.12), we only need to consider the following relationship:

$$\tau < \|\sqrt{\mathbf{W}^{-1}}([\mathbf{I} - (\mathbf{H}+\Delta\mathbf{H})\mathbf{\Lambda}']\mathbf{Hc})\|. \tag{6.13}$$

Our numerical results demonstrate the validity of this claim by showing that the entry values of vector $\sqrt{\mathbf{W}^{-1}}(\mathbf{z}' - \mathbf{H}'\hat{\mathbf{x}}')$ are in the magnitude of 10^{-6} and $\sqrt{\mathbf{W}^{-1}}\{[\mathbf{I} - (\mathbf{H}+\Delta\mathbf{H})\mathbf{\Lambda}']\mathbf{Hc}\}$ are usually in the magnitude 10^{-3} or higher when we randomly construct an *effective* FmDI attack. This, therefore, enables the very existence of a general profile between the minimum $\|\text{diag}(\Delta\mathbf{D})\|$ and the $\mathbf{c}$.

6.3.2.2 Optimization Problem Formulation Under AC Power Flow Model

In addition to the DC power flow model, our objective to minimize the demanded efforts under the constraints of power flow balance laws and D-FACTS capabilities can also be formulated in an AC power flow model, which can be described as

$$\min_{\Delta\mathbf{D}} \quad \|\Delta\mathbf{D}\| \tag{6.14a}$$

$$\text{s.t.} \quad \|\overline{\boldsymbol{\gamma}}'_a(\Delta\mathbf{D})\| > \tau \tag{6.14b}$$

$$d_k^{min} < \Delta d_k < d_k^{max}, \ k \in \mathcal{L} \tag{6.14c}$$

$$P_i = P_{i,G} - P_{i,L} = \sum_{j\in\mathcal{N}_i} P'_{ij}, \ i, j \in \mathcal{N}. \tag{6.14d}$$

It seems the formulas are similar to Eqs. (6.7a)–(6.7d), but it need to be highlighted that the definitions of $\Delta\mathbf{D}$ and $\overline{\boldsymbol{\gamma}}'_a$ under the AC power flow model are different from that under DC model. Concretely, $\mathbf{D} \in \mathbb{R}^{l\times 1}$ is defined as the admittance vector and, therefore, $\Delta\mathbf{D}$ denotes the vector of admittance variations when D-FACTS devices are triggered, i.e.,

$$\Delta\mathbf{D} = (\Delta d_1, \Delta d_2, \cdots, \Delta d_l)^\mathsf{T}. \tag{6.15}$$

Particularly, as for the $\overline{\gamma}'_a$, since in AC power flow model, relationship between the measurement data $\mathbf{z}'_a = \mathbf{z}' + \mathbf{a}$ and system state vector $\mathbf{x_a}'$ is given by

$$\mathbf{z}'_a = \mathbf{z}' + \mathbf{a} = \mathbf{z}' + \mathbf{h}(\mathbf{c}) = \mathbf{h}'(\mathbf{x}'_a) + \boldsymbol{\eta}, \tag{6.16}$$

when FmDI attacks exist and D-FACTS devices are triggered, the normalized measurement residual vector $\overline{\gamma}'_a$ is then expressed as

$$\overline{\gamma}'_a = \sqrt{\mathbf{W}^{-1}}(\mathbf{z}'_a - \hat{\mathbf{z}}'_a) = \sqrt{\mathbf{W}^{-1}}[\mathbf{z}'_a - \mathbf{h}'(\hat{\mathbf{x}}'_a)]. \tag{6.17}$$

The current system states $\hat{\mathbf{x}}'_a$ vector is now estimated by

$$
\begin{aligned}
\hat{\mathbf{x}}'_a &= \min_{\mathbf{x_a}'} [\mathbf{z}'_a - \mathbf{h}'(\mathbf{x}'_a)]^{\mathsf{T}}\mathbf{W}^{-1}[\mathbf{z}'_a - \mathbf{h}'(\mathbf{x}'_a)] \\
&= \min_{\mathbf{x_a}'} [\mathbf{z}' + \mathbf{a} - \mathbf{h}'(\mathbf{x}'_a)]^{\mathsf{T}}\mathbf{W}^{-1}[\mathbf{z}' + \mathbf{a} - \mathbf{h}'(\mathbf{x}'_a)] \\
&= \sum_{i=1}^{m} \frac{(z'_i + h_i(\mathbf{c}) - h'_i(\mathbf{x}'_a))^2}{\sigma_i^2}.
\end{aligned}
\tag{6.18}
$$

Note that the matrix $\mathbf{h}'$ involves the knowledge of vector $\Delta\mathbf{D}$, the relationship of which is highly nonlinear in AC power flow model and is hard to be clearly presented. As we can see the optimization problem to find the minimum efforts by triggering D-FACTS devices to identify FmDI attacks can also be applicable to AC power flow model. Solving the optimization problem in AC power flow model, however, is computationally expensive because this problem is highly nonlinear.

6.3.2.3 Evaluation of the Relationship Between $\|\text{diag}(\Delta\mathbf{D})\|$ and c

In this part, we evaluate the relationship between $\|\text{diag}(\Delta\mathbf{D})\|$ and $\mathbf{c}$ by taking into account all these three classes of FmDI attacks in DC model. It is worth noting that, although our numerical results are collected upon a 7-bus power system (see Fig. 6.5), the method we use to mine the relationship, as shown above, applies to any type of power grids. In this chapter, we solve the optimization problem by considering only activating D-FACTS devices deployed on one branch, which means only one element in $\Delta\mathbf{D}$ is non-zero. In addition, since the values of Δd_k are discrete in that D-FACTS devices are composed of unit modules, the searching space is rather constrained in the range of $[d_k^{min}, d_k^{max}]$. It is, therefore, easy to search all the possible values of Δd_k and further obtain the minimum demanded efforts in a very short time.

First, we consider a case where a single-bus FmDI attack targeted on bus 2 and D-FACTS devices are deployed on branch $(2, 5)$. In this case, we obtain the relationship between the minimum $|\Delta b_{25}|$ and c_2 (see Fig. 6.2) under various

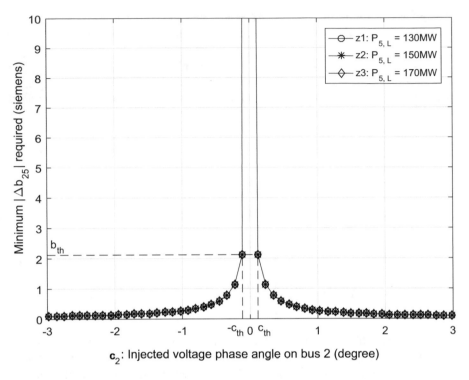

Fig. 6.2 The relationship between the minimum $|\Delta b_{25}|$ and c_2

measurement instants, i.e., $P_{5,L} = 130\,\text{MW}$, $150\,\text{MW}$, and $170\,\text{MW}$, where $|\Delta b_{25}|$ is the absolute susceptance of branch $(2, 5)$ and c_2 is the 2-th element of vector **c**. As can be seen from this figure, the obtain profiles are most likely the same in different measurement instants. This, in turn, justifies the previous claim that the relationship between the minimum $\|\text{diag}(\Delta \mathbf{D})\|$ and **c** is independent of the real-time measurements **z** and system states **x**. Moreover, it can also be seen from each profile that, the greater the absolute value of c_2, the lower minimum efforts are demanded. In other words, it means that it would be easier for power system defenders to identify FmDI attacks with reckless behaviors where large absolute values of **c** are preferred to anticipate huge damages or illicit profits. In addition, when $|c_2| < c_{th}$, either huge efforts are demanded to identify the FmDI attacks or it is actually impossible (beyond the capability of given D-FACTS devices) to identify FmDI attacks using the PFDD approach. Let $c_{th} > 0$ be the threshold of tolerance to voltage phase angle variations, implying the maximum absolute value of injected false voltage phase angle or measurement noise of a voltage phase angle that a power system can tolerate. Specifically, the value of c_{th} can be determined by using Eq. (6.12) with a given τ, and the solutions $\{c_{th}^1, c_{th}^2, \cdots, c_{th}^n\}$ for different buses might be diverse owing to different configurations. As for such cases, c_{th} may take the minimum solution, that is $c_{th} = \min\{c_{th}^1, c_{th}^2, \cdots, c_{th}^n\}$. Correspondingly, given

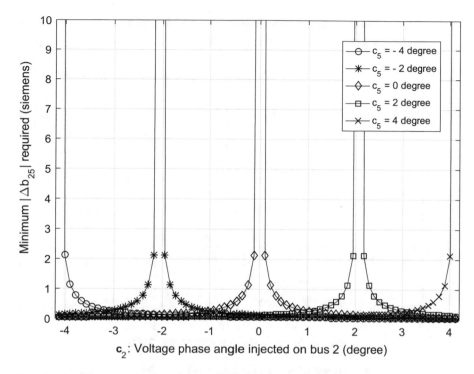

Fig. 6.3 The relationship between the minimum $|\Delta b_{25}|$ and c_2 under various values of c_5

c_{th}, a threshold b_{th} denoting the minimum efforts demanded to identify *effective* FmDI attacks can also be computed as per Eq. (6.12).

In the second case, we consider that an uncoordinated multiple-bus FmDI attack is mounted on buses of both 2 and 5, and D-FACTS devices are deployed on branch (2, 5). Figure 6.3 shows the relationship between the minimum $|\Delta b_{25}|$ and c_2 under diverse values of c_5, the 5-th element of **c**. As we can see that, although with different "central locations," similar profiles to each other and to those shown in Fig. 6.2 are obtained when c_5 varies. In other words, the profile of the minimum efforts demanded for identifying an uncoordinated multiple-bus FmDI attack is analogous to that for a single-bus FmDI attack; however, the exact value is determined by the injected phase difference ($c_2 - c_5$ here) between the two buses, if D-FACTS devices are deployed on the branch in between.

Additionally, Fig. 6.3 also shows some numerical results for cases of detecting a coordinated multiple-bus FmDI attack by using PFDD. For those cases where the injected phase difference between c_2 and c_5 is sufficiently small (less than a tolerance threshold), infinite efforts are required; consequently, PFDD cannot identify such a coordinated multiple-bus FmDI attack by only using the D-FACTS devices deployed on branch (2, 5). Fortunately, we find that as long as any additional branch incident to the super-bus (formed by all the targeted buses in an

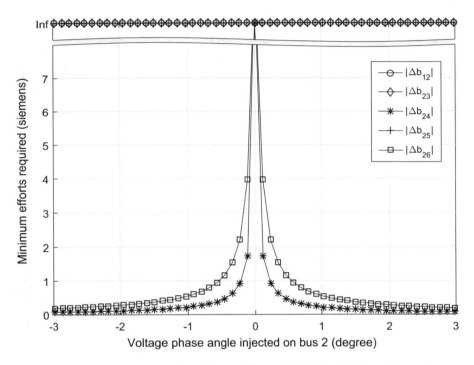

Fig. 6.4 The relationship between the minimum efforts and the injected voltage phase angle

interconnected mode, as aforementioned) is implemented with D-FACTS devices, the coordinated multiple-bus FmDI attacks can still be detected.

This finding is evidenced by the numerical results shown in Fig. 6.4. In the third case, we suppose that D-FACTS devices are implemented on all branches connected to bus 2 and simulate a coordinated multiple-bus FmDI attack on buses $1, 2, 3, 5$, and 7. As Fig. 6.4 shows, only by triggering D-FACTS devices on branches $(2, 4)$ and $(2, 6)$, anomalies can (i.e., FmDI attacks) be observed. The reason behind is that, a same value of voltage phase angle (e.g. $|\theta_a| > c_{th}$) is injected onto all the targeted buses (buses $1, 2, 3, 5$, and 7 here), in a coordinated multiple-bus FmDI attack. In this case, no phase difference can be observed among the above-mentioned coordinated buses. On the contrary, sufficient phase differences can be observed between any pair of un-targeted and targeted buses (say, 4 and 2 or 6 and 2 here). Note that in a special case where all the buses are targeted by a coordinated multiple-bus FmDI attack, no anomaly will be detected by using PFDD due to nonexistence of any uncoordinated bus. However, recall that such an FmDI attack is *ineffective*.

6.3.3   The Minimum Requirement of D-FACTS Devices Implementation to Detect FmDI Attacks

As we can see from the above discussions, it is feasible to identify *effective* FmDI attacks by using the PFDD approach. To further facilitate subsequent discussions, we summarize this finding into Theorem 6.1.

Theorem 6.1 *In the PFDD approach, D-FACTS devices deployed on a branch can be used to identify the existence of effective FmDI attacks targeted on either one or two end bus(es) with degrees both larger than 1 of this branch, as long as the phase angle difference injected by FmDI attackers between these two end buses is greater than a specific tolerance threshold c_{th}.*

Proof According to the definition of the tolerance threshold c_{th}, with a given τ for BDD test, if

$$\mathbf{c} = \underbrace{(0, 0, \cdots, 0, c_{th}, 0, \cdots, 0)}_{n}{}^{\mathsf{T}}, \tag{6.19}$$

then we have

$$\left\| \sqrt{\mathbf{W}^{-1}} \{ \mathbf{z}' - (\mathbf{H} + \Delta\mathbf{H})\hat{\mathbf{x}}' + [\mathbf{I} - (\mathbf{H} + \Delta\mathbf{H})\mathbf{\Lambda}']\mathbf{Hc} \} \right\| \leq \tau, \tag{6.20}$$

according to Eq. (6.12); and for any arbitrarily small positive value Δc, if

$$\mathbf{c} = \underbrace{(0, 0, \cdots, 0, c_{th} + \Delta c, 0, \cdots, 0)}_{n}{}^{\mathsf{T}}, \tag{6.21}$$

then

$$\left\| \sqrt{\mathbf{W}^{-1}} \{ \mathbf{z}' - (\mathbf{H} + \Delta\mathbf{H})\hat{\mathbf{x}}' + [\mathbf{I} - (\mathbf{H} + \Delta\mathbf{H})\mathbf{\Lambda}']\mathbf{Hc} \} \right\| > \tau. \tag{6.22}$$

In this way, let the injected phase angle difference between two ends buses $|c_0| > c_{th}$, and set

$$\mathbf{c} = \underbrace{(0, 0, \cdots, 0, c_0, 0, \cdots, 0)}_{n}{}^{\mathsf{T}}. \tag{6.23}$$

It is easy to obtain the same result as that in Eq. (6.22), which means that in this case the *effective* FmDI attacks can be detected.   □

In this section, we explore the minimum number of branches demanded to be implemented with D-FACTS devices to guarantee a successful detection of all three classes of *effective* FmDI attacks. A theorem shall be proposed showing that the branches installed with D-FACTS devices need to cover at least a spanning tree of the grid graph to ensure detection of *effective* FmDI attacks. Prior to our discussions, we give the following definitions.

Definition 6.3 A branch is defined as a *known branch* if its susceptance (or admittance) value is unchangeable and can be known to the adversaries; otherwise, it is defined as an *unknown branch*.

Usually, we consider a branch implemented with D-FACTS devices as an *unknown branch* in that its susceptance (or admittance) value can be changed by triggering D-FACTS devices. Likewise, a branch without D-FACTS devices being implemented is considered as a *known branch*.

Definition 6.4 A bus is defined as a *protected bus*, if it is connected to at least one *unknown branch*; and an *unprotected bus*, otherwise.

With these definitions in hand, we can prove the following theorem.

Theorem 6.2 *It is feasible to identify effective FmDI attacks targeted on buses or super-buses with degrees larger than 1 by using the PFDD approach, if and only if the unknown branches in a power grid can cover at least a spanning tree of the grid graph.*

Proof Sufficiency We suppose that a line of $n-1$ branches that form a spanning tree $\mathcal{T}$ of a power grid graph $G = \{\mathcal{V}, \mathcal{E}\}$ are implemented with D-FACTS devices. As per Definitions 6.3 and 6.4, these $n-1$ branches are all *unknown branches*. In this case, we know that all buses in this power grid are *protected buses* because any one of them is incident to at least one of these *unknown branches*. As we know, for any type of *effective* single-bus or uncoordinated multiple-bus FmDI attacks, there must be at least one *unknown branch* incident to the targeted bus(es). As per Theorem 6.1, it is feasible for system defenders using the PFDD approach to identify these FmDI attacks with given *unknown branches*. With regard to *effective* coordinated multiple-bus FmDI attacks, at most $n-1$ buses are targeted, leaving at least one bus un-targeted. Thus, there must be a *cut* $C = \{\mathcal{V}^t, \mathcal{V}^u\}$ dividing buses in a grid graph into two sets, namely targeted buses set $\mathcal{V}^t$ and un-targeted buses set $\mathcal{V}^u$, wherein $\mathcal{V}^t \bigcup \mathcal{V}^u = \mathcal{V}$. Importantly, the *cut-set* of C involves edges having one endpoint in $\mathcal{V}^t$ and the other in $\mathcal{V}^u$. If given *unknown branches* covering a spanning tree (see Fig. 6.5a for an example), the *cut-set* must contain at least one *unknown branch* for any type of *effective* coordinated multiple-bus FmDI attacks. With these *unknown branches*, it is feasible to detect *effective* coordinated multiple-bus FmDI attacks by using PFDD (see Theorem 6.1).

Necessity In cases where *unknown branches* in a power grid fail to cover a spanning tree, there must be a *cut* $C = \{\mathcal{V}^1, \mathcal{V}^2\}$ dividing the buses in a grid graph into two sets, i.e., $\mathcal{V}^1$ and $\mathcal{V}^2$, and the *cut-set* of C contains no *unknown branch*. Then a coordinated multiple-bus FmDI attack on all buses in $\mathcal{V}^1$ but none in $\mathcal{V}^2$, or on all buses in $\mathcal{V}^2$ but none in $\mathcal{V}^1$ will not be detected by the PFDD approach. □

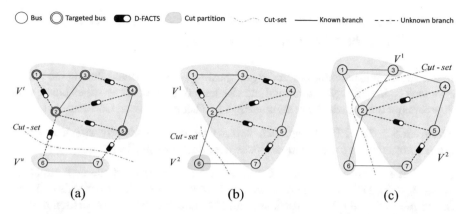

Fig. 6.5 Detection of *effective* FmDI attacks by using PFDD approach under various D-FACTS deployment strategies: (**a**) no *effective* FmDI attack when *unknown branches* contain a spanning tree; (**b**) *effective* single-bus (on $\mathcal{V}^2$) or coordinated multiple-bus (on $\mathcal{V}^1$) FmDI attacks when *unknown branches* fail to contain a spanning tree; and (**c**) *effective* single-bus (on $\mathcal{V}^1$), uncoordinated multiple-bus (on $\mathcal{V}^1$), or coordinated multiple-bus (on $\mathcal{V}^2$) FmDI attacks when *unknown branches* fail to contain a spanning tree—less *unknown branches* compared to (**b**)

Figure 6.5b and c shows examples of FmDI attacks when *unknown branches* fail to contain a spanning tree. In the case as shown in Fig. 6.5b where *unknown branches* fail to contain a spanning tree, buses in partition $\mathcal{V}^1$ can be targeted by *effective* coordinated multiple-bus FmDI attacks and buses in partition $\mathcal{V}^2$ can be targeted by *effective* single-bus FmDI attacks. Both attacks will not be detected. In an even worse case as shown in Fig. 6.5c where fewer *unknown branches* are deployed compared to that in Fig. 6.5b, buses in partition $\mathcal{V}^2$ can be targeted by *effective* coordinated multiple-bus FmDI attacks and buses in partition $\mathcal{V}^1$ can be targeted by either *effective* single-bus or *effective* uncoordinated multiple-bus FmDI attacks. Again, none of these attacks will be detected. It can, therefore, be concluded that *unknown branches* failing to contain a spanning tree in a grid graph leave opportunities for *effective* FmDI attacks to be successful.

6.4 Discussions on the Limitations of PFDD Approach

In this part, we investigate the limitations of using the PFDD approach to identify *effective* FmDI attacks, particularly for those targeted on buses or super-buses with degrees equalling 1.

6.4.1   Limitations of Identifying Effective FmDI Attacks by Using the PFDD Approach

In this subsection, we show our findings regarding the limitations of using the PFDD approach to identify *effective* FmDI attacks targeting on buses or super-buses with degrees equalling 1, which are summarized in Theorem 6.3 and Corollary 6.1.

Theorem 6.3 *The PFDD approach cannot be used to identify effective FmDI attacks targeting on buses or super-buses with degrees equalling 1.*

Proof Let $\epsilon_k \in \{0, 1\}^{l \times 1}$ be a unit column vector where its k-th element equals to 1, and $\delta_i \in \{0, 1\}^{n \times 1}$ be a unit column vector where its i-th element equals to 1. We define $\mu_{ij} \triangleq \delta_i - \delta_j$. Then, matrices $\mathbf{A}$ and $\mathbf{D}$ in Section 6.2 can be written by

$$\mathbf{A} = \sum_{k \in \mathcal{L},\ k \sim \{i,j\}} \epsilon_k \mu_{ij}^\mathsf{T} \text{ and } \mathbf{D} = \sum_{k \in \mathcal{L},\ k \sim \{i,j\}} -b_{ij} \epsilon_k \epsilon_k^\mathsf{T}, \tag{6.24}$$

where $k \sim \{i, j\}$ denotes the branch k connecting buses i and j. Then $\mathbf{DA}$ and $\mathbf{A}^\mathsf{T}\mathbf{DA}$ are, respectively, given by

$$\mathbf{DA} = \sum_{k \in \mathcal{L},\ k \sim \{i,j\}} -b_{ij} \epsilon_k \mu_{ij}^\mathsf{T}, \tag{6.25}$$

and

$$\mathbf{A}^\mathsf{T}\mathbf{DA} = \sum_{k \in \mathcal{L},\ k \sim \{i,j\}} -b_{ij} \mu_{ij} \mu_{ij}^\mathsf{T}. \tag{6.26}$$

Let $\rho_i \in \{0, 1\}^{(n+2l) \times 1}$ be a unit column vector where its i-th entry equals to 1. Based on Eq. (6.2), the $\mathbf{H}$ matrix can be written as

$$\mathbf{H} = \begin{bmatrix} \mathbf{A}^\mathsf{T}\mathbf{DA} \\ \mathbf{DA} \\ -\mathbf{DA} \end{bmatrix} = \begin{bmatrix} \sum_{k \in \mathcal{L},\ k \sim \{i,j\}} -b_{ij} \mu_{ij} \mu_{ij}^\mathsf{T} \\ \sum_{k \in \mathcal{L},\ k \sim \{i,j\}} -b_{ij} \epsilon_k \mu_{ij}^\mathsf{T} \\ \sum_{k \in \mathcal{L},\ k \sim \{i,j\}} b_{ij} \epsilon_k \mu_{ij}^\mathsf{T} \end{bmatrix}$$

$$= \sum_{k \in \mathcal{L},\ k \sim \{i,j\}} -b_{ij}(\rho_i - \rho_j + \rho_{n+k} - \rho_{n+l+k}) \mu_{ij}^\mathsf{T}. \tag{6.27}$$

For a Single-Bus with Degree 1 Given that bus $\zeta \in \mathcal{N}$ with degree equalling 1 is targeted by an *effective* single-bus FmDI attack, and bus $\gamma \in \mathcal{N}$ is the only neighbor of bus ζ connected by branch $\ell \in \mathcal{L}$. With an aim of injecting θ_a to bus ζ, the attacker designs

$$\mathbf{c} = (\underbrace{0, 0, \cdots, 0, \overbrace{\theta_a}^{\zeta\text{-th}}, 0, \cdots, 0)}_{n}^{\mathsf{T}}, \tag{6.28}$$

which can then be rewritten as $\mathbf{c} = \theta_a \boldsymbol{\delta}_\zeta$. In this way, the attack vector $\mathbf{a}$ can be written by

$$\begin{aligned}
\mathbf{a} = \mathbf{Hc} &= -b_{\zeta\gamma}(\boldsymbol{\rho}_\zeta - \boldsymbol{\rho}_\gamma + \boldsymbol{\rho}_{n+\ell} - \boldsymbol{\rho}_{n+l+\ell})\boldsymbol{\mu}_{\zeta\gamma}^{\mathsf{T}}\theta_a\boldsymbol{\delta}_\zeta \\
&= -b_{\zeta\gamma}\theta_a(\boldsymbol{\rho}_\zeta - \boldsymbol{\rho}_\gamma + \boldsymbol{\rho}_{n+\ell} - \boldsymbol{\rho}_{n+l+\ell})(\boldsymbol{\delta}_\zeta^{\mathsf{T}} - \boldsymbol{\delta}_\gamma^{\mathsf{T}})\boldsymbol{\delta}_\zeta \\
&= -b_{\zeta\gamma}\theta_a(\boldsymbol{\rho}_\zeta - \boldsymbol{\rho}_\gamma + \boldsymbol{\rho}_{n+\ell} - \boldsymbol{\rho}_{n+l+\ell}).
\end{aligned} \tag{6.29}$$

If D-FACTS devices implemented on branch ℓ are triggered, the susceptance value of this branch is changed to $b'_{\zeta\gamma}$ and the $\mathbf{H}$ matrix is correspondingly changed to $\mathbf{H}'$. Then, we can derive the following important finding:

$$\mathbf{a} = \mathbf{Hc} = \mathbf{H}'\mathbf{c}' = -b'_{\zeta\gamma}\theta'_a(\boldsymbol{\rho}_\zeta - \boldsymbol{\rho}_\gamma + \boldsymbol{\rho}_{n+\ell} - \boldsymbol{\rho}_{n+l+\ell}), \tag{6.30}$$

where

$$\mathbf{c}' = (\underbrace{0, 0, \cdots, 0, \overbrace{\theta'_a}^{\zeta\text{-th}}, 0, \cdots, 0)}_{n}^{\mathsf{T}}, \text{ and } \theta'_a = \frac{b_{\zeta\gamma}\theta_a}{b'_{\zeta\gamma}}. \tag{6.31}$$

Based on Eqs. (6.6) and (6.11), $\|\overline{\boldsymbol{\gamma}}'_a(\Delta\mathbf{D})\|$ can be further written as

$$\begin{aligned}
\|\overline{\boldsymbol{\gamma}}'_a(\Delta\mathbf{D})\| &= \|\sqrt{\mathbf{W}^{-1}}(\mathbf{z}' - \mathbf{H}'\hat{\mathbf{x}}' + \mathbf{a} - \mathbf{H}'\Delta\mathbf{x})\| \\
&= \|\sqrt{\mathbf{W}^{-1}}(\mathbf{z}' - \mathbf{H}'\hat{\mathbf{x}}' + \mathbf{H}'\mathbf{c}' - \mathbf{H}'\boldsymbol{\Lambda}'\mathbf{H}'\mathbf{c}')\| \\
&\underset{\boldsymbol{\Lambda}'\mathbf{H}'=\mathbf{I}}{=\!=\!=\!=} \|\sqrt{\mathbf{W}^{-1}}(\mathbf{z}' - \mathbf{H}'\hat{\mathbf{x}}' + \mathbf{H}'\mathbf{c}' - \mathbf{H}'\mathbf{c}')\| \\
&= \|\sqrt{\mathbf{W}^{-1}}(\mathbf{z}' - \mathbf{H}'\hat{\mathbf{x}}')\| < \tau.
\end{aligned} \tag{6.32}$$

As we can see, no FmDI alarm shall be issued if using the PFDD approach to identify *effective* FmDI attacks targeting on single-buses with degrees equalling 1.

For a Super-Bus with Degree 1 Given that an *effective* coordinated multiple-bus FmDI attack is targeted at a set of buses $\mathcal{S} = \{\zeta, \zeta + 1, \cdots, \zeta + t\}$, where t is a positive integer. These buses can form into a super-bus with its degree equalling 1, where branch ℓ is the only external branch to this super-bus that connects buses from ζ to γ. With an aim of injecting θ_a to this super-bus, the attacker designs

$$\mathbf{c} = (0, 0, \cdots, 0, \overbrace{\theta_a}^{\zeta\text{-th}}, \overbrace{\theta_a}^{(\zeta+1)\text{-th}}, \cdots, \overbrace{\theta_a}^{(\zeta+t)\text{-th}}, 0, \cdots, 0)^{\mathsf{T}}, \qquad (6.33)$$

which can be rewritten as $\mathbf{c} = \theta_a \sum_{t=0}^{t} \delta_{\zeta+t}$. Correspondingly, the attack vector $\mathbf{a}$ is rewritten as

$$\mathbf{a} = \mathbf{H}\mathbf{c} = \left(\sum_{k\in\mathcal{L}^{\mathcal{S}},\, k\sim\{i,j\}} -b_{ij}(\boldsymbol{\rho}_i - \boldsymbol{\rho}_j + \boldsymbol{\rho}_{n+k} - \boldsymbol{\rho}_{n+l+k})\boldsymbol{\mu}_{ij}^{\mathsf{T}} \right) \times \left(\theta_a \sum_{t=0}^{t} \delta_{\zeta+t} \right)$$

$$= \left(\sum_{k\in\mathcal{L}^{\mathcal{S}},\, k\sim\{i,j\}} -b_{ij}\theta_a(\boldsymbol{\rho}_i - \boldsymbol{\rho}_j + \boldsymbol{\rho}_{n+k} - \boldsymbol{\rho}_{n+l+k})(\delta_i^{\mathsf{T}} - \delta_j^{\mathsf{T}}) \right) \times \left(\sum_{t=0}^{t} \delta_{\zeta+t} \right),$$
$$(6.34)$$

where $\mathcal{L}^{\mathcal{S}}$ denotes the branches connecting to any of the buses in set $\mathcal{S}$. Note that $\forall i, j$ satisfying $k \sim \{i, j\}$ and $k \in \mathcal{L}^{\mathcal{S}}$, there must have $i, j \in \mathcal{S}$ except for one case where $i = \zeta$ and $j = \gamma$. In this way, Eq. (6.34) can be rewritten as

$$\mathbf{a} = \left(\sum_{k\in\mathcal{L}^{\mathcal{S}},\, k\sim\{i,j\}} -b_{ij}\theta_a(\boldsymbol{\rho}_i - \boldsymbol{\rho}_j + \boldsymbol{\rho}_{n+k} - \boldsymbol{\rho}_{n+l+k}) \times (\delta_i^{\mathsf{T}} - \delta_j^{\mathsf{T}}) \right) \left(\sum_{t=0}^{t} \delta_{\zeta+t} \right)$$

$$= \left(\sum_{k\in\{\mathcal{L}^{\mathcal{S}}\setminus\ell\},\, k\sim\{i,j\}} -b_{ij}\theta_a(\boldsymbol{\rho}_i - \boldsymbol{\rho}_j + \boldsymbol{\rho}_{n+k} - \boldsymbol{\rho}_{n+l+k}) \times (\delta_i^{\mathsf{T}} - \delta_j^{\mathsf{T}})(\delta_i + \delta_j) \right)$$

$$+ \left(\sum_{k=\ell,\, \ell\sim\{\zeta,\gamma\}} -b_{\zeta\gamma}\theta_a \times (\boldsymbol{\rho}_\zeta - \boldsymbol{\rho}_\gamma + \boldsymbol{\rho}_{n+\ell} - \boldsymbol{\rho}_{n+l+\ell}) \times (\delta_\zeta^{\mathsf{T}} - \delta_\gamma^{\mathsf{T}})\delta_\zeta \right)$$

$$= \left(\sum_{k\in\{\mathcal{L}^{\mathcal{S}}\setminus\ell\},\, k\sim\{i,j\}} -b_{ij}\theta_a(\boldsymbol{\rho}_i - \boldsymbol{\rho}_j + \boldsymbol{\rho}_{n+k} - \boldsymbol{\rho}_{n+l+k}) \times 0 \right)$$

$$+ \left(-b_{\zeta\gamma}\theta_a(\boldsymbol{\rho}_\zeta - \boldsymbol{\rho}_\gamma + \boldsymbol{\rho}_{n+\ell} - \boldsymbol{\rho}_{n+l+\ell}) \times 1 \right)$$

$$= -b_{\zeta\gamma}\theta_a(\boldsymbol{\rho}_\zeta - \boldsymbol{\rho}_\gamma + \boldsymbol{\rho}_{n+\ell} - \boldsymbol{\rho}_{n+l+\ell}).$$
$$(6.35)$$

We have the same conclusion as that shown in Eq. (6.29). Hence, we see that for a super-bus with degree equalling 1, if targeted by an *effective* coordinated multiple-bus FmDI attack, we still have $\mathbf{H}\mathbf{c} = \mathbf{H}'\mathbf{c}'$, leading to the failure of detecting such an

FmDI attack using the PFDD approach. It is worth noting that although this FmDI attack remains undetected, it is equivalent to another FmDI attack with

$$\mathbf{c}' = (\underbrace{0, 0, \cdots, 0, \overbrace{\theta'_a}^{\zeta\text{-th}}, \overbrace{\theta'_a}^{(\zeta+1)\text{-th}}, \cdots, \overbrace{\theta'_a}^{(\zeta+t)\text{-th}}, 0, \cdots, 0)}_{n}^{\mathsf{T}}, \tag{6.36}$$

where $\theta'_a = b_{\zeta\gamma}\theta_a/b'_{\zeta\gamma}$ when the susceptance value of branch $\ell \sim \{\zeta, \gamma\}$ is $b'_{\zeta\gamma}$. □

Corollary 6.1 *Given a super-bus or a single-bus ζ with degree equalling 1 (as to a super-bus, ζ denotes the bus connecting the only external branch), the external connected bus γ, and the branch ℓ connecting these ζ and γ. Without knowing the susceptance value of branch ℓ, as long as FmDI attackers are able to inject P_a to P_ζ, $-P_a$ to P_γ, and P_a to $P_{\zeta\gamma}$, the PFDD approach cannot be used to identify such an FmDI attack, where P_ζ, P_γ, $P_{\zeta\gamma}$, and P_a, respectively, represent the total nodal power injection value of bus ζ, the total nodal power injection value of bus γ, the power flow of branch $\ell \sim \{\zeta, \gamma\}$, and a constant power value that the attacker wish to inject.*

Proof As we know, $\mathbf{z}$ is an $m \times 1 = (n + 2l) \times 1$ column vector involving n nodal power injection values $\mathbf{P_I} = \{P_1, P_2, \cdots, P_n\}$, $2l$ power flows, that is $\mathbf{P_F} = \{P_{ij}|i, j \in N, k \sim \{i, j\}, k \in \mathcal{L}\}$ and $-\mathbf{P_F} = \{-P_{ij}|i, j \in N, k \sim \{i, j\}, k \in \mathcal{L}\}$. We can rewrite $\mathbf{z}$ by

$$\mathbf{z} = (\mathbf{P_I}, \mathbf{P_F}, -\mathbf{P_F})^{\mathsf{T}} = \sum_{i=1}^{n} P_i \boldsymbol{\rho}_i + \sum_{k \in \mathcal{L}, \, k \sim \{i,j\}} P_{ij} \boldsymbol{\rho}_{n+k} + \sum_{k \in \mathcal{L}, \, k \sim \{i,j\}} -P_{ij} \boldsymbol{\rho}_{n+l+k}. \tag{6.37}$$

If FmDI attackers can inject P_a to P_ζ, $-P_a$ to P_γ, and P_a to $P_{\zeta\gamma}$, it means that the FmDI attacker is able to construct an attack vector

$$\mathbf{a} = P_a(\boldsymbol{\rho}_\zeta - \boldsymbol{\rho}_\gamma + \boldsymbol{\rho}_{n+\ell} - \boldsymbol{\rho}_{n+l+\ell}). \tag{6.38}$$

Based on Eq. (6.30), we can rewrite $\mathbf{a}$ as

$$\mathbf{a} = \mathbf{Hc} = \mathbf{H}'\mathbf{c}' = P_a(\boldsymbol{\rho}_\zeta - \boldsymbol{\rho}_\gamma + \boldsymbol{\rho}_{n+\ell} - \boldsymbol{\rho}_{n+l+\ell}). \tag{6.39}$$

If D-FACTS devices are triggered by using PFDD and $b_{\zeta\gamma}$ is updated to $b'_{\zeta\gamma}$,

$$\mathbf{c}' = (\underbrace{0, 0, \cdots, 0, \overbrace{\theta'_a}^{\zeta\text{-th}}, 0, \cdots, 0)}_{n}^{\mathsf{T}}, \text{ and } \theta'_a = \frac{P_a}{b'_{\zeta\gamma}} \tag{6.40}$$

for a single-bus ζ with degree equalling 1, and

$$\mathbf{c}' = (0, 0, \cdots, 0, \overbrace{\theta'_a}^{\zeta\text{-th}}, \overbrace{\theta'_a}^{(\zeta+1)\text{-th}}, \cdots, \overbrace{\theta'_a}^{(\zeta+t)\text{-th}}, 0, \cdots, 0)^{\mathsf{T}}, \qquad (6.41)$$

$$\underbrace{}_{n}$$

for a super-bus ζ with degree equalling 1, containing buses $\zeta, \zeta + 1, \cdots$, and $\zeta + t$. According to Eq. (6.32), we see that such an FmDI attack targeted on either a single-bus or a super-bus with degree 1 cannot be detected by using PFDD. It is worth noting that though such FmDI attacks would be successful even without the knowledge of $b_{\zeta\gamma}$, the attackers under such cases however shall have no idea of the real injected phase angle offset θ'_a, because they cannot immediately obtain the value of $b'_{\zeta\gamma}$ after D-FACTS devices are activated. In other words, they do not know how much impact they can cause or how many profits they can obtain through a successful FmDI attack.   □

6.4.2   Case Studies

In this subsection, we take 8-bus and 39-bus power systems (see Figs. 6.6 and 6.7) as examples to illustrate *effective* FmDI attacks targeting on a single-bus and a super-bus with degree 1, respectively.

6.4.2.1   Case 1: An Effective FmDI Attack Targeted on Bus 1 in an 8-Bus Power System

If an FmDI attacker would like to inject θ_a towards bus 1 phase angle θ_1, first he constructs a vector $\mathbf{c}$ by

$$\mathbf{c} = \theta_a \boldsymbol{\delta}_1 = (\theta_a, 0, 0, 0, 0, 0, 0, 0)^{\mathsf{T}}, \qquad (6.42)$$

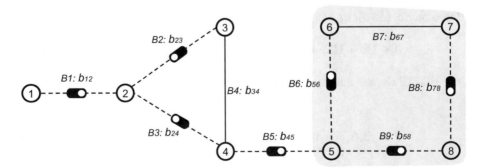

Fig. 6.6 An illustrative 8-bus power grid with D-FACTS implementation covering a spanning tree

as well as an attack vector **a** by

$$\mathbf{a} = \mathbf{Hc} = -b_{12}\theta_a(\boldsymbol{\rho}_1 - \boldsymbol{\rho}_2 + \boldsymbol{\rho}_{n+1} - \boldsymbol{\rho}_{n+l+1}). \tag{6.43}$$

Then, the FmDI attacker falsifies the measurement data $\mathbf{z}$ by $\hat{\mathbf{z}} = \mathbf{z} + \mathbf{a}$. In this way, it is equivalent to adding $-b_{12}\theta_a$ to P_1, $b_{12}\theta_a$ to P_2, $-b_{12}\theta_a$ to P_{12}, and $b_{12}\theta_a$ to P_{21} through compromised meters. If system defenders activate D-FACTS devices implemented on branch $B1$, b_{12} is updated to b'_{12}. According to Eqs. (6.30) and (6.32), this FmDI attack cannot be identified by using the PFDD approach, however, an offset of $\theta'_a = b_{12}\theta_a/b'_{12}$ other than θ_a is injected to θ_1. It is equivalent to an FmDI attack with a vector

$$\mathbf{c} = (\theta'_a, 0, 0, 0, 0, 0, 0, 0)^\mathsf{T}, \tag{6.44}$$

while the susceptance value of branch $B1$ is b'_{12}.

6.4.2.2 Case 2: An Effective FmDI Attack Targeted on a Super-Bus Composed of Buses 5, 6, 7, and 8 in an 8-Bus Power System

If an FmDI attacker would like to inject θ_a to the phase angles of a super-bus comprising buses 5, 6, 7, and 8, first he constructs a vector **c** by

$$\mathbf{c} = \theta_a(\delta_5 + \delta_6 + \delta_7 + \delta_8) = (0, 0, 0, 0, \theta_a, \theta_a, \theta_a, \theta_a)^\mathsf{T}, \tag{6.45}$$

as well as an attack vector **a** by

$$\mathbf{a} = \mathbf{Hc} = -b_{45}\theta_a(\boldsymbol{\rho}_5 - \boldsymbol{\rho}_4 + \boldsymbol{\rho}_{n+5} - \boldsymbol{\rho}_{n+l+5}). \tag{6.46}$$

Then, the FmDI attacker falsifies the measurement data $\mathbf{z}$ by $\hat{\mathbf{z}} = \mathbf{z} + \mathbf{a}$. In this way, it is equivalent to adding $-b_{45}\theta_a$ to P_5, $b_{45}\theta_a$ to P_4, $-b_{45}\theta_a$ to P_{54}, and $b_{45}\theta_a$ to P_{45} through compromised meters. If system defenders activate the D-FACTS devices implemented on branch $B5$, b_{45} is updated to b'_{45}. As per Eqs. (6.35) and (6.32), this FmDI attack cannot be identified by using the PFDD approach, but an offset of $\theta'_a = b_{45}\theta_a/b'_{45}$ is injected to θ_5, θ_6, θ_7, and θ_8, respectively. It is equivalent to an FmDI attack with a vector

$$\mathbf{c} = (0, 0, 0, 0, \theta'_a, \theta'_a, \theta'_a, \theta'_a)^\mathsf{T} \tag{6.47}$$

when the susceptance value of branch $B5$ is b'_{45}.

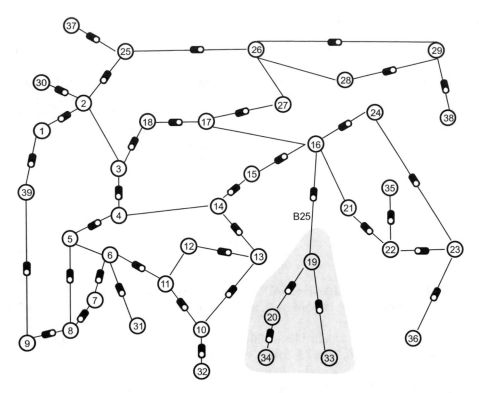

Fig. 6.7 An illustrative 39-bus power grid with D-FACTS implementation covering a spanning tree

6.4.2.3   Case 3: An Effective FmDI Attack Targeted on a Super-Bus Composed of Buses 19, 20, 33, and 34 in the IEEE 39-Bus Power System

If an FmDI attacker would like to inject θ_a to the phase angles of a super-bus comprising buses 19, 20, 33, and 34, first he constructs a vector **c** by

$$\mathbf{c} = \theta_a(\delta_{19} + \delta_{20} + \delta_{33} + \delta_{34}) = (0, 0, \cdots, 0, \overbrace{\theta_a}^{19\text{-th}}, \overbrace{\theta_a}^{20\text{-th}}, 0, \cdots, 0, \overbrace{\theta_a}^{33\text{-th}}, \overbrace{\theta_a}^{34\text{-th}})^{\mathsf{T}},$$
$$\underbrace{}_{39}$$

(6.48)

as well as an attack vector **a** by

$$\mathbf{a} = \mathbf{Hc} = -b_{16,19}\theta_a(\boldsymbol{\rho}_{19} - \boldsymbol{\rho}_{16} + \boldsymbol{\rho}_{n+25} - \boldsymbol{\rho}_{n+l+25}),$$

(6.49)

where 25 is the index of branch $B25$ connecting buses 16 and 19. Then, the attacker falsifies the measurement data $\mathbf{z}$ by $\hat{\mathbf{z}} = \mathbf{z} + \mathbf{a}$. In this way, it is equivalent to adding $-b_{16,19}\theta_a$ to P_{19}, $b_{16,19}\theta_a$ to P_{16}, $-b_{16,19}\theta_a$ to $P_{19,16}$, and $b_{16,19}\theta_a$ to $P_{19,16}$

through compromised meters. If system defenders trigger the D-FACTS devices implemented on branch $B25$, $b_{16,19}$ is updated to $b'_{16,19}$. As per Eqs. (6.35) and (6.32), this FmDI attack cannot be identified by using the PFDD approach, but an offset of $\theta'_a = b_{16,19}\theta_a/b'_{16,19}$ is injected to $\theta_{19}, \theta_{20}, \theta_{33}$, and θ_{34}, respectively. It is equivalent to an FmDI attack with a vector

$$\mathbf{c} = \underbrace{(0, 0, \cdots, 0, \overbrace{\theta'_a}^{19\text{-th}}, \overbrace{\theta'_a}^{20\text{-th}}, 0, \cdots, 0, \overbrace{\theta'_a}^{33\text{-th}}, \overbrace{\theta'_a}^{34\text{-th}})^{\mathsf{T}}}_{39}, \tag{6.50}$$

when the susceptance value of branch $B25$ is $b'_{16,19}$.

6.5 Summary

In this chapter, we thoroughly explored the feasibility and limitations of using the PFDD approach to identify FmDI attacks on smart grids. By considering three classes of FmDI attacks, that is, single-bus, uncoordinated multiple-bus, and coordinated multiple-bus FmDI attacks, we obtained the respective profiles of the minimum efforts demanded for using the D-FACTS devices to identify FmDI attacks. Importantly, we also strictly proved that the PFDD approach can guarantee the detection of the existence of all three classes of FmDI attacks, if and only if the implementation of D-FACTS devices can cover branches containing at least a spanning tree of the power grid graph. Last, the limitations of using the PFDD approach were carefully studied, with findings that the PFDD approach cannot be used to identify effective FmDI attacks targeting on buses or super-buses with degrees equalling 1. Further studies are needed to investigate whether PFDD by activating D-FACTS devices can cause potential instability or other hidden problems in power grids, and how often to activate D-FACTS devices would be an optimal solution to detect FmDI attacks in smart grids.

References

1. Li, B., Lu, R., Wang, W., & Choo, K.-K. R. (2016). DDOA: A Dirichlet-based detection scheme for opportunistic attacks in smart grid cyber-physical system. *IEEE Transactions on Information Forensics and Security, 11*(11), 2415–2425.
2. Li, B., Lu, R., Wang, W., & Choo, K.-K. R. (2017). Distributed host-based collaborative detection for false data injection attacks in smart grid cyber-physical system. *Journal of Parallel and Distributed Computing, 103*, 32–41.
3. Liu, Y., Ning, P., & Reiter, M. K. (2011). False data injection attacks against state estimation in electric power grids. *ACM Transactions on Information and System Security, 14*(1), 13.

4. Rahman, M. A., & Mohsenian-Rad, H. (2012). False data injection attacks with incomplete information against smart power grids. In *Proceedings of the IEEE GLOBECOM, Anaheim, CA, USA, December 3–7*, 2012 (pp. 3153–3158).
5. Deng, R., Xiao, G., & Lu, R. (2017). Defending against false data injection attacks on power system state estimation. *IEEE Transactions on Industrial Informatics, 13*(1), 198–207.
6. Valenzuela, J., Wang, J., & Bissinger, N. (2013). Real-time intrusion detection in power system operations. *IEEE Transactions on Power Systems, 28*(2), 1052–1062.
7. Che, L., Liu, X., & Li, Z. (2018). Fast screening of high-risk lines under false data injection attacks. *IEEE Transactions on Smart Grid, 10*(4), 4003–4014.
8. Che, L., Liu, X., Li, Z., Shuai, Z., Li, Z., & Wen, Y. (2018). Mitigating false data attacks induced overloads using a corrective dispatch scheme. *IEEE Transactions on Smart Grid, 10*(3), 3081–3091.
9. Kovacs, E. (2017). WikiLeaks releases details on CIA hacking tools. *SecurityWeek*, March 2017 [Online]. Available: http://www.securityweek.com/wikileaks-releases-details-cia-hacking-tools
10. Deng, R., & Liang, H. (2018). False data injection attacks with limited susceptance information and new countermeasures in smart grid. *IEEE Transactions on Industrial Informatics, 15*(3), 1619–1628.
11. Kim, T. T., & Poor, H. V. (2011). Strategic protection against data injection attacks on power grids. *IEEE Transactions on Smart Grid, 2*(2), 326–333.
12. Morrow, K. L., Heine, E., Rogers, K. M., Bobba, R. B., & Overbye, T. J. (2012). Topology perturbation for detecting malicious data injection. In *Proceedings of the 45th Hawaii International Conference on System Science (HICSS), Maui, HI, USA, January 4–7*, 2012 (pp. 2104–2113).
13. Rahman, M. A., Al-Shaer, E., & Bobba, R. B. (2014). Moving target defense for hardening the security of the power system state estimation. In *Proceedings of the First ACM Workshop on Moving Target Defense, Scottsdale, AZ, USA, November 07*, 2014 (pp. 59–68).
14. Tian, J., Tan, R., Guan, X., & Liu, T. (2017). Hidden moving target defense in smart grids. In *Proceedings of the 2nd Workshop on Cyber-Physical Security and Resilience in Smart Grids, Pittsburgh, PA, USA, April 18–21*, 2017 (pp. 21–26).
15. Divan, D., & Johal, H. (2007). Distributed FACTS - A new concept for realizing grid power flow control. *IEEE Transactions on Power Electronics, 22*(6), 2253–2260.
16. Tian, J., Tan, R., Guan, X., & Liu, T. (2018). Enhanced hidden moving target defense in smart grids. *IEEE Transactions on Smart Grid, 10*(2), 2208–2223.
17. Rogers, K., & Overbye, T. (2008). Some applications of distributed flexible AC transmission system (D-FACTS) devices in power systems. In *Proceedings of the 40th North American Power Symposium (NAPS), Calgary, AB, Canada, September 28–30*, 2008 (pp. 1–8).
18. Sou, K. C., Sandberg, H., & Johansson, K. H. (2011). Electric power network security analysis via minimum cut relaxation. In *Proceedings of the 50th Conference on Decision and Control and European Control Conference (CDC-ECC), Orlando, FL, USA, December 12–15*, 2011 (pp. 4054–4059).

Chapter 7
Epilogue

In this chapter, we summarize our contributions of this monograph, and propose some potential research directions for future work.

7.1 Summary of Contributions

The research focuses in this monograph lie in main topics relating to FDI attacks in smart grid CPSs including modeling and impact evaluation of FDI attacks, novel detection approaches for FDI attacks—both FmDI and FcDI attacks. Specifically, our main research contributions are summarized as follows:

- In Chap. 3, we have developed a stochastic Petri net based analytical model to evaluate and analyze the system reliability of smart grid CPSs, specifically against topology attacks under system countermeasures (i.e., intrusion detection systems and malfunction recovery techniques). Topology attacks are evolved from FDI attacks, where attackers initialize FDI attacks by tempering with both measurement data and grid topology information. This analytical model is featured by bolstering both transient and steady-state analyses of system reliability.
- In Chap. 4, we have proposed a distributed host-based collaborative detection scheme to detect FmDI attacks in smart grid CPSs. It is considered in this chapter that the PMUs, deployed to measure the operating states of power grids, can be compromised by FmDI attackers, and the trusted HMs assigned to each PMU are employed to monitor and assess PMUs' behaviors. Neighboring HMs make use of the majority voting algorithm based on a set of predefined normal behavior rules to identify the existence of abnormal measurement data collected by PMUs. In addition, an innovative reputation system with an adaptive reputation updating

© Springer Nature Switzerland AG 2020

B. Li et al., *Detection of False Data Injection Attacks in Smart Grid Cyber-Physical Systems*, Wireless Networks,
https://doi.org/10.1007/978-3-030-58672-0_7

algorithm is also designed to evaluate the overall operating status of PMUs, by which FmDI attacks as well as the attackers can be distinctly observed.

- In Chap. 5, we have proposed a Dirichlet-based detection scheme for FcDI attacks in hierarchical smart grid CPSs. In the future hierarchical paradigm of a smart grid CPS, it is considered that the decentralized LAs responsible for local management and control can be compromised by FcDI attackers. By issuing fake or biased commands, the attackers anticipate to manipulate the regional electricity prices with the purpose of illicit financial gains. The proposed scheme builds a Dirichlet-based probabilistic model to assess the reputation levels of LAs. This probabilistic model, used in conjunction with a designed adaptive reputation incentive mechanism, enables quick and efficient detection of FcDI attacks as well as the attackers.

- In Chap. 6, we have systematically explored the feasibility and limitations of detecting FmDI attacks in smart grid CPSs using D-FACTS devices. Recent studies have investigated the possibilities of proactively detecting FmDI attacks on smart grid CPSs by using D-FACTS devices—the PFDD approach. In this chapter, the feasibility of using PFDD to detect FmDI attacks is investigated by considering single-bus, uncoordinated multiple-bus, and coordinated multiple-bus FmDI attacks, respectively. It is proved that PFDD can detect all these three types of FmDI attacks targeted on buses or super-buses with degrees larger than 1, as long as the deployment of D-FACTS devices covers branches at least containing a spanning tree of the grid graph. The minimum efforts required for activating D-FACTS devices to detect each type of FmDI attacks are, respectively, evaluated. In addition, the limitations of this approach are also discussed, and it is strictly proved that PFDD is not able to detect FmDI attacks targeted on buses or super-buses with degrees equalling 1.

7.2 Future Work

Our studies have made significant progress in enhancing smart grid security and reliability, especially in detection and mitigation of FDI attacks. There are still plenty of work that can be done along the same direction. The following research topics will be investigated in the future as a continuation of my Ph.D. monograph.

- *Secure state estimation and proactive mitigation of FDI attacks*: Liu et al. mentioned that if the attackers can intrude into the control systems and obtain the knowledge of **H** matrix, the state estimation can be bypassed leading to a success of FDI attacks. As we see, there is still a need to ensure secure state estimation even though the attackers are strong enough to get access to the control systems. As we introduced in Sect. 2.1, the control center needs to calculate the **H** matrix and use it to conduct state estimation. With a closer look at Eqs. (2.6) and (2.37), we notice that to achieve state estimation and bad data detection, the control center can choose to use matrices $\mathbf{\Lambda}$ and $\mathbf{\Omega} \triangleq \mathbf{I} - \mathbf{H\Lambda}$, respectively, instead of

H matrix itself. This makes it possible to hide **H** matrix behind **Λ** and **Ω**. Since **H** matrix is the critical information for attackers to construct FmDI attacks, it may therefore be a useful way to mitigate FmDI attacks by hiding **H** matrix when conducting state estimation and bad data detection. We are motivated to conduct future research studies to achieve secure estimation and bad data detection without explicit content of **H** matrix, e.g., using applied cryptography. Unlike traditional approaches for passively detecting FmDI attacks, this approach can be considered as a proactive way to mitigate FDI attacks in smart grid CPS, which would be a breakthrough for solving such a problem and shed lights for upcoming studies.

- *Investigation of side effects by using D-FACTS devices*: In Chap. 6, we have discussed the feasibility and limitations of using PFDD to detect FmDI attacks in smart grid. However, there may be concerns about the potential negative impacts on power systems caused by using D-FACTS devices. For example, activating D-FACTS devices may, to some extent, compromise the optimal power flows and, therefore, lead to certain power losses. If it is true, is there any suboptimal status for power flows that we may transfer to when activating D-FACTS devices. It may also be a concern that whether activating D-FACTS devices can cause transient instabilities or any other hidden problem to the electric grid. If yes, how to contain such problems remains open, which deserves extensive investigations in the future.
- *Prototyping and real-world implementation of proposed solutions*: In order to promote security and reliability of real smart grid infrastructures, we are planning to deploy the solutions presented in this monograph over a test bed, and evaluate and refine the prototypes in a real-world implementation. We expect to cooperate with those teams who have a test bed for smart grid CPS, and test all our proposed solutions.

Index

© Springer Nature Switzerland AG 2020
B. Li et al., *Detection of False Data Injection Attacks in Smart Grid Cyber-Physical Systems*, Wireless Networks,
https://doi.org/10.1007/978-3-030-58672-0

Printed in the United States
by Baker & Taylor Publisher Services